Synthesis Lectures on Emerging Engineering Technologies

This series publishes short books on current engineering technologies that are gaining prominence, as well as promising technologies that are being developed, for an audience of researchers, advanced students, engineers and other professionals, and entrepreneurs.

Maria Mathew · Rony Rajan Paul ·
Ann Rose Abraham · A. K. Haghi

Electrospun Porous Nanofibers

An Introduction

Maria Mathew
Department of Chemistry
National Institute of Technology Calicut
Calicut, Kerala, India

Ann Rose Abraham
Department of Physics
Sacred Heart College
Thevara, Kerala, India

Rony Rajan Paul
Department of Chemistry
CMS College Kottayam (Autonomous)
Kottayam, Kerala, India

A. K. Haghi
Institute of Molecular Chemistry
Coimbra University
Coimbra, Portugal

ISSN 2381-1412   ISSN 2381-1439 (electronic)
Synthesis Lectures on Emerging Engineering Technologies
ISBN 978-3-031-86105-5   ISBN 978-3-031-86106-2 (eBook)
https://doi.org/10.1007/978-3-031-86106-2

This Springer imprint is published by the registered company Springer Nature Switzerland AG
The registered company address is: Gewerbestrasse 11, 6330 Cham, Switzerland

If disposing of this product, please recycle the paper.

Contents

About the Authors

Maria Mathew is currently a research scholar at the Materials Research Laboratory, National Institute of Technology Calicut (NITC). Her research focuses on the development of electrospun fibrous membranes for environmental applications, with a strong emphasis on advanced nanomaterials and sustainable technologies. She has made notable contributions to the field, including the publication of research papers and book chapters in the area of nanotechnology. Her broader research interests encompass nanomaterial synthesis, environmental remediation, and the design of functional materials for diverse applications.

Rony Rajan Paul received his Ph.D. from CSIR-NIIST, Thiruvananthapuram, India, working with Prof. Vijay Nair. He has been a visiting fellow at the Universities of Potsdam and Duesseldorf, Germany, and Nottingham Trent University, UK. After a post-doctoral stay at KU Leuven, Belgium, he returned to his alma mater CMS College, Kottayam, as an assistant professor. In 2024, he was awarded SERB-SIRE fellowship for research stay at KU Leuven. He has published extensively in international journals of good repute and has an h-index 13.

Ann Rose Abraham, Ph.D. is currently an assistant professor of Physics at Sacred Heart College (Autonomous), Thevara, Cochin, India. Her academic journey has been marked by notable achievements and conducting ongoing research at the Materials Research Laboratory (MRL), Sacred Heart College, Kochi. Dr. Abraham received her M.Sc., M.Phil., and Ph.D. degrees in Physics from the School of Pure and Applied Physics, Mahatma Gandhi University, Kerala, India. She has authored, co-authored, edited, or co-edited more than 60 publications, including books, book chapters, and papers in peer-reviewed journals. She is a reviewer of many international journals. She has several publications to her credit in many peer-reviewed high impact journals of international repute, such as Elsevier, Taylor and Francis, Springer, *Journal of Physical Chemistry C*, *Physical Chemistry Chemical Physics*, *New Journal of Chemistry*, and *Philosophical Magazine*. She has research experience at various national institutes including Bose Institute, SAHA Institute of Nuclear Physics, UGC-DAE CSR Centre, Kolkata and has collaborations with various international laboratories, such as the Université de Lorraine (France), University of Johannesburg, and Institute of Physics Belgrade. She is the recipient of young researcher award in physics, a prestigious forum to showcase intellectual capability. She has expertise in the field of materials science, nanomagnetic materials, multiferroics, polymeric nanocomposites, biomaterials, etc.

A. K. Haghi is a retired professor and has written, co-written, edited, or co-edited more than 1000 publications, including books, book chapters, and papers in refereed journals with over 4200 citations and h-index of 34, according to the Google Scholar database. Professor Haghi holds a B.Sc. in urban and environmental engineering from the University of North Carolina (USA) and holds two M.Sc. degrees, one in mechanical engineering from North Carolina State University (USA) and another one in applied mechanics, acoustics, and materials from the Université de Technologie de Compiègne (France). He was awarded a Ph.D. in engineering sciences at Université de Franche-Comté (France).

Professor Haghi's extensive educational background and supervisory roles underscore his expertise and contributions to the field of engineering sciences. He is appointed as an honorary research associate (HRA) at University of Coimbra, Portugal. He is a regular reviewer of leading international journals.

Introduction to Electrospinning of Nanofibers

Introduction

Electrospinning is one of the most adaptable and unique technique used for the development of synthetic non-woven fibers by applying high energy electrical potential [1]. Electrospinning allows the facile development of nanofibers with diameters in the nanometer range. Electrospinning is preferred over other conventional techniques like solvent casting and phase separation [2]. Electrospun nanofibers are characterized by several interesting properties such as high surface area (SA) to volume (V) ratio, inherent porous structure, high degree of interconnections and easily tunable surface properties. The fundamental principle of this technique is the formation of nanofibers by the uniaxial stretching of visco-elastic polymer solution by the effect of an electrostatic force. Nanofibers have recently attained significant research interest due to its versatility across various sectors such as energy storage, tissue engineering, catalysis, drug delivery, sensors and more [3–5]. Electrospinning allows the fabrication of nanostructures with materials ranging from natural to synthetic polymers, even ceramics, metals and also other bioactive components can be introduced into fibers [6].

The central focus of this chapter is to give an introduction into the electrospinning technique. The basic working principle behind the electrospinning technique is that a polymer solution is introduced into a needle charged with high voltage to form fine fibrous structures. During the electrospinning process various parameters like applied solution viscosity, voltage, concentration, flow rate, temperature and humidity can influence the fiber morphology and diameter [7, 8]. In this chapter, we discuss the influence of these parameters collectively called 'electrospinning parameters. Different types of electrospinning techniques are also briefly included [9] in the chapter.

© The Author(s), under exclusive license to Springer Nature Switzerland AG 2025
M. Mathew et al., *Electrospun Porous Nanofibers*, Synthesis Lectures on Emerging
Engineering Technologies, https://doi.org/10.1007/978-3-031-86106-2_1

Historical Background

The webs constructed by silk worms and spiders were the nature inspiration for the production of man-made fibers. By then human beings started fabricating many fibers from both natural (wool and cotton) and synthetic (polyester) sources. Nylon was the first synthetic fiber that gained huge attention and extended the scope of synthetic fibers for various applications [10].

Several methods were developed to fabricate synthetic fibers from polymers including dry spinning, wet spinning, melt spinning and gel spinning. In all these methods, the polymer jet was formed under an external shear force when passing through the spinnerets and fibers are solidified either by precipitation or drying. One of the greatest disadvantages of these methods is their limited stretching, due to which the fiber diameter cannot reach the nanoscale range.

In 1600, William Gilbert was the first person who observed the deformation of spherical droplet into cone shape on interaction with a high voltage electric field. By eighteenth century Stephen Gray comes with his observation that the expulsion of tiny jets from the tip of liquid droplet during this process is due to atomisation of water droplets in presence of electric field. The first known electrospraying experiment, in which water was sprayed into an aerosol in an electric field, was carried out in 1747 by Abbe Nollet [11]. After a long gap, Charles V. Boys proved in 1887 that fibres may be formed from a viscoelastic liquid when exposed to an external electric field [12]. He employed an insulated tiny dish connected to a high voltage power source to generate fibers from different viscous liquids like beeswax, shellac and collodion. The development of continuous fibres of nanometres to micrometres size ranges was made possible by this innovation, which is now often referred to as "electrospinning".

The first patent on developing a prototype of the electrospinning setup was submitted by John Francis Cooley and also William Morton in the early 1900s [13]. In a commercial scale, it was in the Soviet Union the electrospun fibers were first utilized for the development of air filters for capturing aerosol particles. Since then, several research works have been conducted to study how the spherical droplet deforms into conical shape under the impact of electric field. It was Geoffrey Taylor who described the formation of 'Taylor cone' by the deformation of spherical droplet when the electric field strength is increased beyond a critical value [14]. In the early1990s, the research group headed by Gregory Rutledge and Darrell Reneker established that different organic polymeric materials could be used for development of electrospun nanofibers and their findings gave new life to electrospinning. Then gradually electrospinning emerged as the preferred method for the fabrication of nanoscale fibers. Till now the research on developing new electrospinning technique is extensively going on for enhancing its utility for numerous applications including biosensors, filtration, energy harvesting, and storage applications.

Basic Principle of Electrospinning

A basic electrospinning system includes three main components: (1) a high voltage power source, (2) a syringe pump system that includes a needle, and (3) a grounded collector. Figure 1.1 [15] shows a schematic illustration of fundamental electrospinning technique. Initially, the electrospinning polymer solution have to be prepared in appropriate solvent. The prepared polymer solution has to be taken in a syringe comprising a needle. This syringe is now clamped in syringe pump system. A high voltage is then applied across the edge of needle and collector which charges polymer solution. It causes a charge repulsion in the electrospinning solution. As time progresses, the solution extrudes out from the needle tip as a hemispherical droplet. Once the voltage is sufficient enough to resist the surface tension, the hemispherical drops transform into a cone shaped structure called '*Taylor cone*'. Any more rise in the applied voltage leads to expulsion of charged jet of fluid from the Taylor cone. The ejected liquid jet is attracted towards the collector. Solidified nanofibers are gathered over the collector as the solvent evaporates throughout the flight between the needle's end and the collector [16].

Electrospinning Parameters

The diameter and morphology of the fibrous membrane has great significance in its application. Fibers possessing narrow fiber diameter, uniform, and non-beaded structures are always preferred. For obtaining nanofibers with very fine structural features, several factors have to be optimized. These factors are collectively known as electrospinning parameters (Fig. 1.2). These include processing factors (flow rate, distance between tip of

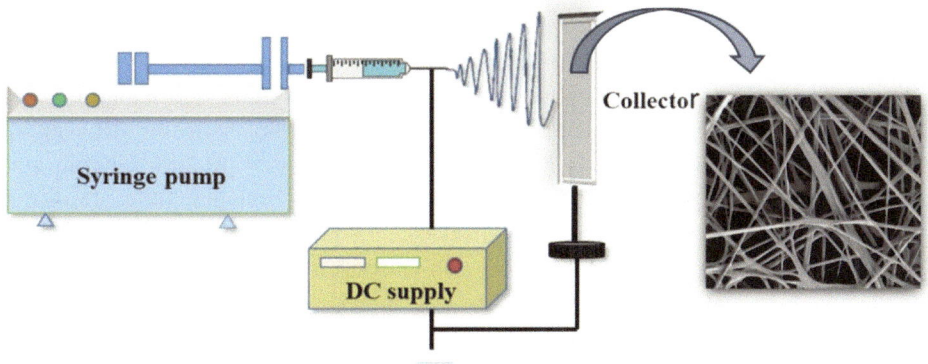

Fig. 1.1 Schematic representation of basic electrospinning setup

Factors affecting electrospinning process

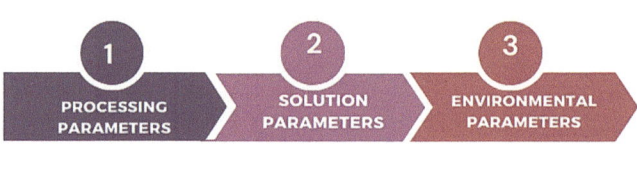

- Applied Voltage
- Flow Rate
- Needle tip to collector distance

- Concentration
- Molecular weight
- Viscosity
- Conductivity/Surface charge density

- Temperature
- Humidity

Fig. 1.2 Electrospinning parameters

needle and collector and applied voltage), solution parameters (viscosity, concentration, molecular weight, conductivity) and ambient parameters (humidity and temperature) [17].

Processing Parameters

Applied Voltage

One of the significant factors that influence electrospinning process is applied voltage. Generally, current that flow from the voltage system to polymeric solution via tip of needle is responsible for the deformation of liquid droplet to Taylor cone. Only when this applied voltage rise high above a critical value the charged polymer jet ejects from the needle tip to form ultrathin fibers. With further rise in voltage, the electrostatic force of repulsion within the polymer increases causing the stretching of polymer jet to form fibers with smaller diameters [18]. Also, several research groups pointed out that, higher voltages facilitate the development of fibers with wider fiber diameter.

Liu et al. [19] reported the impact of voltage on the nanofiber morphology and jet number in the spinning of PVA nanofibers. Table 1.1 shows the change in diameter with voltage. Initially, the fiber diameter reduced as voltage increased till 20 kV, but however the diameter of fibres increased at 25 and 30 kV. At low voltages, only a single jet is formed at the droplet tip while at higher voltages more than one jet is generated. This resulted in higher fiber diameter with non-uniform morphology. Dietzal et al. [20] identified the formation of beaded nanofibers with increased applied voltage in PEO/water system.

Rate of Flow

Flow rate is defined as the amount of the polymeric solution ejected from the tip of needle at unit time. Flow rate must be effectively controlled for the formation of uniform,

Table 1.1 Variation of fiber diameter with applied voltage in the production of PVA nanofibers

Voltage (kV)	Diameter (nm)
7	442.7 ± 59.8
10	272.8 ± 35.7
15	232.9 ± 29.0
20	159.3 ± 23.6
25	186.7 ± 43.4
30	213.6 ± 64.9

beadless fibers with smaller diameters. All polymeric solutions will have a critical flow rate and any flow rate above or below this critical value lead to the development of asymmetrical fibers with broader fiber diameter [21]. Generally, at lower flow rates the solvent will have ample time period for evaporation forming thin fibres, while, higher flow rates generate beads and other defects. A flow rate of 0.5 ml/hr was determined to be the ideal rate in a study by Zargham et al. [22] to comprehend the impact of flow rate on the fibre shape of Nylon-6 as it formed a stable Taylor cone with the smallest variation of fibre diameters. The rate of flow was adjusted from 0.1 to 1.5 ml/hr. When the rate of flow is high above 0.5 ml/hr, higher amounts of polymeric solution were ejected forming bigger droplets, which leads to breakages in the polymer jet, due to the gravitational force effect.

Needle Tip-to-Collector Distance

The needle tip-to-collector distance often referred as the working distance plays a critical part in influencing the morphology of electrospun nanofiber. In general, shorter working distance reduces the time of flight and fibers will not get enough time for solvent evaporation. This leads to beaded nanofibers. On the other hand, extending the distance provides sufficient time for evaporation of solvent and also extending the polymer jet resulting in the formation of narrow fibers [23]. Certain other factors like deposition time and Rayleigh instability are influenced by this working distance, hence, an optimum distance has to being maintained for the generation of smooth fibers.

Solution Parameters

Polymer Concentration and Viscosity

Concentration of the polymer is an imperative aspect that controls the extending of charged jet [24]. Polymer concentration also affects the viscosity and surface tension of the solution [21]. If the concentration is too low, the intertwined polymeric chains breakdown into fragments culminating in beaded fibers. Increase in concentration increases the viscosity of solution. Solution viscosity indicates the extent to which the polymeric chains are entangled in a solution. So, as concentration is increased above a critical value, the chain entanglement increases and the solution will not exit the tip of needle. The impact of polymer content on the mechanical characteristics and morphology of cellulose acetate and poly(vinyl chloride) (PVC) nanofiber mats was investigated by Tarus et al. [25]. For nanofibers of cellulose acetate at low concentration (7, 10 and 10%), beaded fibers were formed. In case of nanofiber mats of PVC at 12% concentration, beaded fibers were obtained and at 14 and 16% concentration smooth fibers were attained. These changes are directly related to the polymer concentration, which influences the solution viscosity. In electrospinning, a low viscosity solution has low viscoelastic force which prevents the proper stretching of the electrospinning jet. This causes the fragmentation of polymer jets which leads to bead formation [26].

Molecular Weight

Another important element that affects the fiber size and morphology is molecular weight [27]. Generally, thicker fibres are formed when molecular weight increases. Koski et al. [28] examined the effect of molecular weight on the morphology or shape of electrospun polyvinyl alcohol (PVA) fibers by spinning PVA of different molecular weights. For molecular weight ranging from 9000 to 13,000 g/mol, circular fibers with bead-on-string arrangement was obtained with an average fiber diameter between 250 nm and 1 μm. With increase in molecular weight (31,000–50,000) flat fibers with diameter in micrometer was observed. Here PVA is a hydrophilic molecule. Hence, with increase in molecular weight the solution viscosity increases due to its high degree of hydrolysis. Due to the gel structure of PVA in water, solvent evaporation rate reduces as molecular weight increases. In such cases, wet fibers get collected over the collector. The reason for observing flat fibers at higher molecular weight can be attributed to this. In contrast to this, at low molecular weight solvent evaporates rapidly to give thin fibers with circular cross section.

Conductivity

Another element influencing the fibre morphology and Taylor cone production is the conductivity of the polymeric solutions. Conductivity of the polymer solution is responsible for the charge build up within polymer surface on applied electric field, which leads to stretching and elongation of the polymer jet [29]. No charges will form at the droplet's surface to create a Taylor cone in a fluid with reduced conductivity, which does not lead to electrospinning. Conversely, an increased solution conductivity facilitates formation of stable Taylor cone resulting in the formation of uniform fibers. A polymer solutions conductivity can be enhanced by adding of conductive additives like salts or metal nanoparticles [21]. To improve the conductivity of polymers with intrinsic microporosity (PIMs), Topuz et al. [30] incorporated ammonium salts to the solution and studied their effect on fiber morphology. Electrospinning of PIMs with 7.5% of tetraethylammonium bromide (TEAB) enabled the development of fibers which are bead-free, however, in the absence of TEAB droplet splashing, beaded fibers were observed. The salt addition influences the electrospinning in two different ways: (i) Salt addition boost the ion count in the polymer, increasing polymer's surface charge density and also electrostatic force on application of external electric field (ii) it lead to increase in conductivity of polymers, causing reduction in electric field at the solution surface [31].

Environmental Parameters

Temperature

One another important factor is the processing temperature [32]. However, there are only few reports on the influence of temperature on the morphology and diameter of fiber. Any variations in the working temperature can affect the nanofiber formation [33]. Generally, an increase in temperature from the ambient conditions decreases surface tension and viscosity of polymer leading to the formation of uniform, smooth fibers.

At higher temperatures polymer molecules have more freedom to move that decreases the viscosity of solution. The lower solution viscosity provides more stretching rate and consequently thinning of electrospun fibers. When the temperature is lowered, the solvent evaporation rate reduces and solidification of jet of polymer takes more time. In this case, due to the slow solvent evaporation the stretching continues leading to small fiber diameter [34]. This is confirmed by Clerck et al. [35] in his investigation on formation of nanofibers using poly(vinylpyrrolidone) (PVP) and cellulose acetate. The morphology of the fibers was analyzed at three different ambient temperatures (283, 293, and 303 K). In the case of PVP nanofibers, the average fiber diameter at 283 and 303 K was smaller than at 293 K. This was due to the two opposing effects discussed above. At lower temperature (283 K) solvent evaporation rate is the predominant factor and at higher temperature (303 K) the lower viscosity is the dominant factor. However, with cellulose acetate only at temperature 303 K at a relative humidity of 45 and 60% a complete nonwoven mat was obtained.

Humidity

Relative humidity (RH) plays a decisive role in pore formation on nanofibers [36]. In investigation by Raksa et al. [37] on the effect of RH on surface structure and mechanical properties of silk fibroin blended PVA nanofibers, they obtained smooth beadless fibers at a RH of 80%. At higher relative humidities, atmosphere holds more water molecules and as a result the solidification takes longer time. As the solidification time increases, the exposure of polymer jet to voltage induced stretching also increases which reduces the fiber diameter. Another interesting feature is the pore formation on the surface and interiors of fibers by controlling RH. Porous structures can be induced on nanofibers through various phase separation methods [38 39]. Many of the mechanisms are driven by RH. One among them is the vapor induced phase separation (VIPS) mechanism. In VIPS mechanism porosity can be induced by spinning with a solvent miscible with water (DMF) that possess a higher boiling point (b.p) within a humid atmosphere. Since the

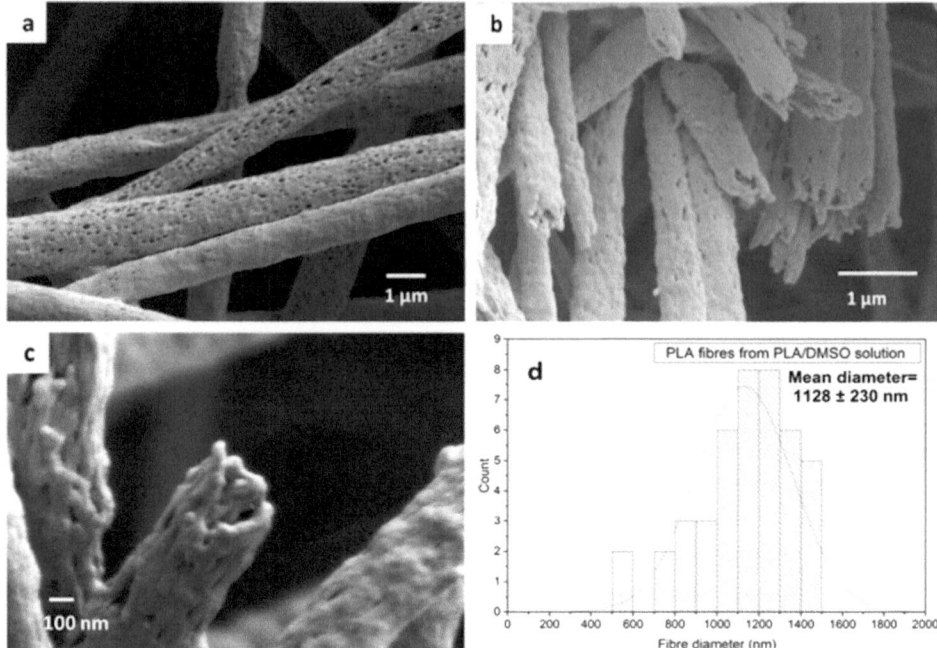

Fig. 1.3 SEM images of porous polylactic acid (PLA) fibers formed by VIPS mechanism in a humid environment (RH 55%); **a, b** surface porosity **c** internal porosity **d** fiber diameter distribution. Reprinted with permission from Ref. [40] Copyright Elsevier 2018

solvent is highly miscible with water, it will absorb water vapor and it penetrates into fiber jet, causing a phase separation. Once the solvent and the water vapor completely evaporate, it leaves pores on the fiber surface and interiors. The SEM images of porous PLA nanofibers fabricated via VIPS mechanism using dimethyl sulfoxide (DMSO) as the water miscible solvent [40] are shown in Fig. 1.3.

Different Modes of Electrospinning

Instead of the single needle-based basic electrospinning technique, many other variations of electrospinning are also known. Mainly, there are two classifications of electrospinning depending on the setup for electrospinning: needle-based and needleless electrospinning [16].

Needle-Based Electrospinning

As the name suggests, in needle-based electrospinning setup a needle-like spinneret is used for electrospinning the solution. The formation of a steady "Taylor cone" is an indicator of a stable electrospinning process. Some common examples for needle-based electrospinning involve monoaxial, co-axial and multi spinneret.

Monoaxial Electrospinning

Monoaxial (uniaxial) electrospinning is considered a popular electrospinning technique used for the fabrication of nanofibers using a conventional single nozzle. Figure 1.1 represents a typical monoaxial electrospinning technique. As already discussed in Sect. 1.2, a high electric filed is applied during the process between the needle tip and the metallic collector. Under the influence of electric field, the polymer solution is transformed in to fibers with nano diameters. The main advantage of monoaxial electrospinning is its simple fabrication procedure. Hence, it has been widely used in drug delivery applications [41]. High drug loading capacity, sustained release, high encapsulation efficiency, ease of operation are some of the features of monoaxial electrospinning [42].

Coaxial Electrospinning

In coaxial electrospinning, a coaxial needle containing two concentric and hollow needles are used to yield nanofibers with core-sheath structure (Fig. 1.4) [43]. Coaxial electrospinning permits the selectivity of an extensive range of spinning materials, especially materials with poor spinnability. The sheath and core precursor solutions are fed through the outer and inner nozzle respectively [44]. The electrical conductivity of sheath and core solution determines the charge distribution. Conceptually, the basic working principle of coaxial electrospinning is similar to monoaxial electrospinning. When sufficient voltage is applied, charge accumulates on the sheath solution coming out through the outer nozzle [45]. Charge repulsion causes the sheath solution to stretch, just like in monoaxial electrospinning. The sheath polymer jet emerges from the Taylor cone when the voltage hits a threshold. The tension produced creates a viscous drag force on the core solution during the process inside the sheath solution [46]. The core solution thus gets transformed into conical structure and carried along with the sheath solution to form core-sheath or core–shell nanofibers. The relative alignment of the outer and inner nozzle is an important factor that affects the fiber formation [47]. Figure 1.5 represents the TEM image (a) and SEM image (b) of PVP-TiO$_2$ core-sheath fibers fabricated via coaxial electrospinning [48].

The polymer jet may undergo whipping and bending instabilities while in flight which may also impact the resulting fiber morphology. Another important factor is the optimization of the properties of core and sheath solution. From the various studies conducted on the solution properties (viscosity, conductivity, miscibility, vapor pressure and surface tension), some of the findings are concluded below.

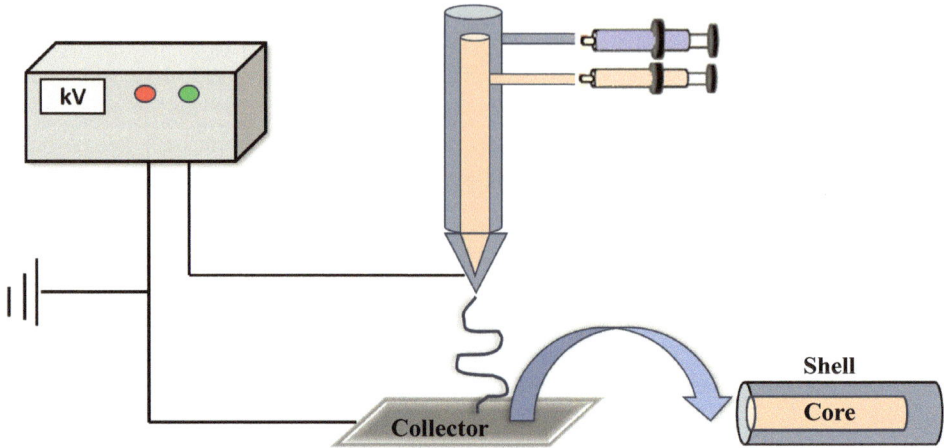

Fig. 1.4 Schematic representation of coaxial electrospinning setup

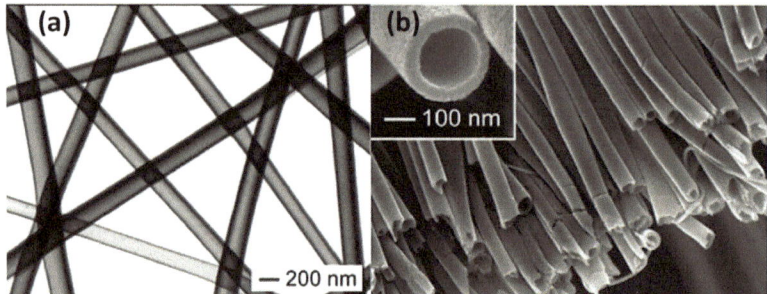

Fig. 1.5 Core-sheath PVP-TiO_2 hollow fibers fabricated via coaxial electrospinning **a** TEM image, **b** SEM image. Copyright © 2004 American Chemical Society

1. The sheath solution must be an electrospinnable polymer solution.
2. Viscosity of sheath solution must be higher than core solution.
3. Sheath solution must possess higher conductivity.
4. Solvents of low vapor pressure preferred.
5. Minimal surface tension across the interface between the sheath and core solutions.

Therefore, the so-produced core–shell nanofibers, as they may be used for drug encapsulation [49] effectively, have been widely exploited in biomedical applications such as tissue engineering, drug delivery, and wound healing [50].

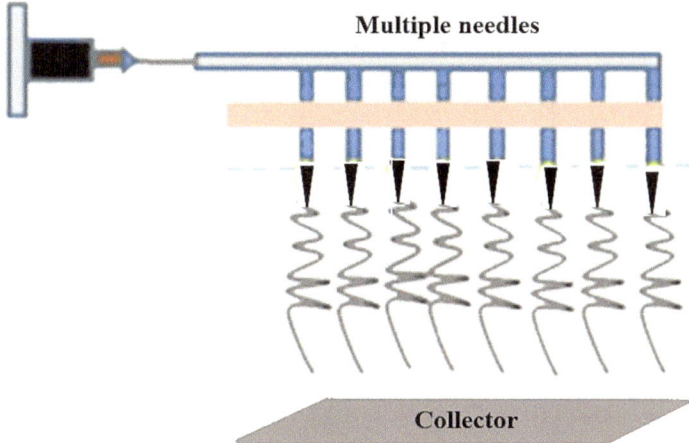

Fig. 1.6 Multi-needle electrospinning

Multi-needle Electrospinning

Multi-needle electrospinning allows the nanofiber mass production by using a greater number of needles [51]. Figure 1.6 represents a multi-needle electrospinning setup. The technology used in multi-needle electrospinning is analogous to that in single-needle electrospinning, however, the needles are arranged according to a set of rules to complete the process of electrospinning. The aim of this setup is to increase the nanofiber yield by increasing the number of polymer jets. There are certain factors that need to be considered in multi-needle electrospinning. One is the needle arrangement. Tomaszewski et al. [52]. constructed three different needle arrangements: linear, elliptic and concentric and compared their utility by spinning polyvinyl alcohol (PVA)/water system. Similarly, several studies have been conducted on the needle arrangement [53, 54]. Another factor is the electric field strength [55]. Electric field can affect the morphology and fiber diameter to a great extent. Since multiple needles are used in this process, a uniform distribution of electric field is indispensable for thin and uniform fiber formation. Yang et al. [54] used a shielding ring to overcome this issue. The main limitation of multi-needle electrospinning is the repulsion from adjacent polymer jets causing the difficulty in the proper collection of nanofibers [56].

Needleless Electrospinning

Needleless electrospinning emerged as a modification to overcome the limitations of needle-based electrospinning such as clogging and low fiber production [57]. In this

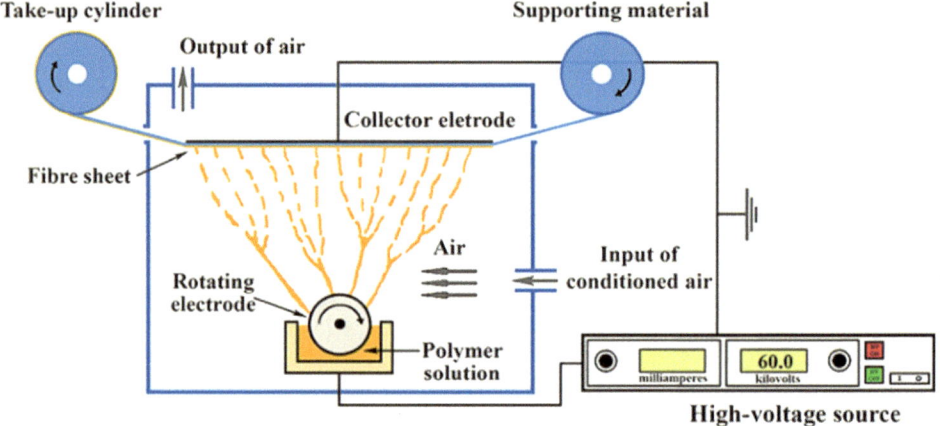

Fig. 1.7 Schematic representation of needless electrospinning [58]

approach, nanofibers are fabricated from an open liquid surface. Needleless electrospinning is a self-organizing process that occurs in free surface, where nanofibers are produced by inducing instability in polymer solution by applied electrostatic forces. This leads to the elongation and thinning of droplets to form polymer jet from the Taylor cone formed at the apex of the solution [58] (Fig. 1.7).

The history of needleless electrospinning begun with the use of annular electrode as the spinning nozzle by Simm et al. in 1979 [59]. Later in 2004, magnetic field was used to prompt spike formations on the solution surface to initiate the electrospinning process [60]. In 2005, the first patented needless electrospinning setup was commercialized with a brand name 'Nanospider' by Kotek et al. [61] This system consisted of a rotating cylinder as the spinneret, which was partially immersed and rotated in the polymeric solution to develop a thin layer. Then electrospinning process was initiated by applying the electric field. Therefore, in needless electrospinning, spinnerets/nozzles play an important role in fiber formation and quality.

Applications of Electrospun Nanofibrous Membranes

Electrospun nanofibrous membranes offer a wide range applications like energy storage, sensing, catalysis, biomedical, and environmental treatments. Figure 1.8 portrays the applications of electrospun nanofibrous membranes. The high specific surface area and porosity of electrospun nanofibrous membranes makes it useful as filtration membranes for air and water purification. The highly porous structure facilitates the transport of gas or liquid molecules without much resistance, thus effectively removing pollutants including toxic ions, heavy metals, particulate matters and other organic pollutants (dyes, pesticides, plasticizers) [63]. Optimising the pore size and fibre diameter of nanofiber membranes can improve their filtration efficacy when compared to more conventional filtration materials like activated carbon. Another exciting application of electrospun nanofibers are energy harvesting and storage. The nanofiber membranes made from ceramics, polymers and carbon can act as good supports for enzyme and metal nanoparticle-based catalysis [64, 65]. For the effective harvesting, storage and conversion of energy many devices such as rechargeable batteries, solar cells, supercapacitors and fuel cells were developed. Electrodes are one of the major components that allows the electrons or ion conduction for the proper functioning of these devices. Nanofiber electrodes can offer many advantages including large electrode/electrolyte contact area, efficient electron/ion transfer, lightweight and flexibility to the devices. All these features attract nanofiber-based devices for energy applications [66, 67]. Over past few decades, electrospun nanofibers have received immense applications in biomedical field. Nanofibers developed by electrospinning can mimic the morphology of native extracellular matrix, revealing their importance in biomedical applications [68]. Through engineering of their structural properties such as porosity, diameter, mechanical properties, surface functional groups and biodegradability, flexible structures can be fabricated. These structures enhance the cell adhesion, proliferation, and differentiation. Besides, electrospun membranes offers diverse approaches for carrying drugs such as coating and encapsulation [69]. Drug encapsulated nanofibers provides a great environment to treat infections, controlled release, skin regeneration, and reduced time for wound healing process [70]. Other fields of applications include defense and security, porous textiles, and food packaging.

Limitations and Challenges

Despite all the advantages discussed, electrospinning technique also have some limitations. One of the most recognized limitations of conventional electrospinning is its low productivity, which may not be suitable for applications requiring large quantities of materials. Some other disadvantages include insufficient mechanical strength, hydrophobic nature, non-uniformity in fiber diameter and morphology, and limited material solubility. The safety concerns originating from electrospinning is mainly related to the toxicity of the polymers and solvents being used during the process. The use of toxic organic solvents that can endanger human beings and living beings which makes this technology less environmentally friendly [71]. Many studies have revealed the harmful effects of inhaling relatively short nanofibers. It was reported that, injection of electrospun Ag-nanofibers of 5–20 µm length into the mice lungs caused severe respiratory problems [72]. Although electrospinning is a highly effective technology with a wide scope of applications, these are some of the issues that need to be addressed to enhance their acceptability for commercialization.

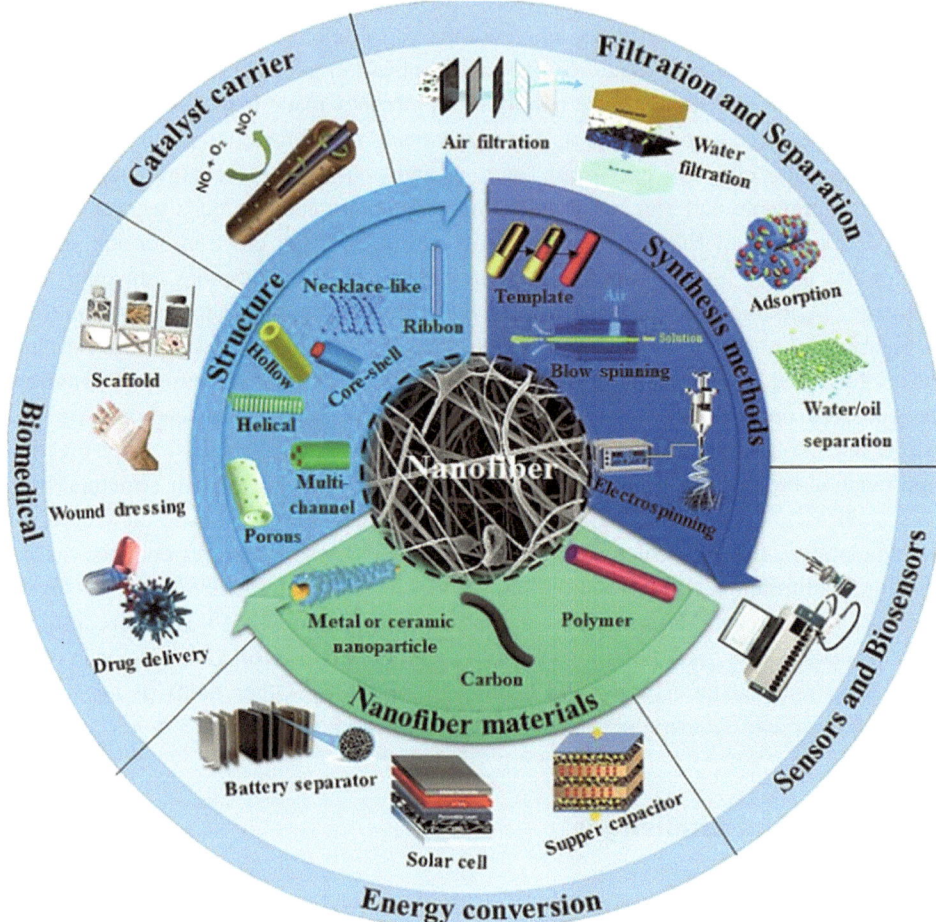

Fig. 1.8 Schematic representation of various applications of electrospun nanofibers. Reprinted with permission from Ref. [62]. Copyright Elsevier 2024

References

1. Xue, J., Wu, T., Dai, Y. & Xia, Y. Electrospinning and electrospun nanofibers: Methods, materials, and applications. *Chem. Rev.* **119**, 5298–5415 (2019).
2. Yongquan, D., Ming, W., Lin, C. & Mingjun, L. Preparation, characterization of P(VDF-HFP)/[bmim]BF4 ionic liquids hybrid membranes and their pervaporation performance for ethyl acetate recovery from water. *Desalination* **295**, 53–60 (2012).
3. Liu, Q., Zhu, J., Zhang, L. & Qiu, Y. Recent advances in energy materials by electrospinning. *Renew. Sustain. Energy Rev.* **81**, 1825–1858 (2018).

4. Liu, Y. *et al.* A review on recent advances in application of electrospun nanofiber materials as biosensors. *Curr. Opin. Biomed. Eng.* **13**, 174–189 (2020).

5. Rahmati, M. *et al.* Electrospinning for tissue engineering applications. *Prog. Mater. Sci.* **117**, 100721 (2021).

6. Ahmadi Bonakdar, M. & Rodrigue, D. Electrospinning: Processes, Structures, and Materials. *Macromol* **4**, 58–103 (2024).

7. Rodoplu, D. & Mutlu, M. Effects of electrospinning setup and process parameters on nanofiber morphology intended for the modification of quartz crystal microbalance surfaces. *J. Eng. Fiber. Fabr.* **7**, 118–123 (2012).

8. Au, H. T., Pham, L. N., Vu, T. H. T. & Park, J. S. Fabrication of an antibacterial non-woven mat of a poly(lactic acid)/chitosan blend by electrospinning. *Macromol. Res.* **20**, 51–58 (2012).

9. Blachowicz, T. & Ehrmann, A. Most recent developments in electrospun magnetic nanofibers: A review. *J. Eng. Fiber. Fabr.* **15**, 1558925019900843 (2020).

10. McIntyre, J. E. *Synthetic Fibres: Nylon, Polyester, Acrylic, Polyolefin.* (2004).

11. Guan, B. & Cole, R. B. The Background to Electrospray. in *The Encyclopedia of Mass Spectrometry* (eds. Gross, M. L. & Caprioli, R. M.) 132–140 (Elsevier, Boston, 2016). https://doi.org/10.1016/B978-0-08-043848-1.00016-X.

12. Ghosal, K., Agatemor, C., Tucker, N. & Kny, E. Electrical Spinning to. *Electr. Spinn. to Electrospinning a Br. Hist.* 1–23 (2018) https://doi.org/10.1039/9781788011006-00001.

13. Sarkar, K. *et al.* Electrospinning to Forcespinning^{TM}. *Mater. Today* **13**, 12–14 (2010).

14. Keirouz, A. *et al.* The History of Electrospinning: Past, Present, and Future Developments. *Adv. Mater. Technol.* **8**, 2201723 (2023).

15. P. S. Suja C. R. Reshmi, P. S. & Sujith, A. Electrospun Nanofibrous Membranes for Water Purification. *Polym. Rev.* **57**, 467–504 (2017).

16. Sharma, G. K. & James, N. R. Electrospinning: The Technique and Applications. in *Recent Developments in Nanofibers Research* (eds. Khan, M. & Chelladurai, S. J. S.) (IntechOpen, Rijeka, 2022). https://doi.org/10.5772/intechopen.105804.

17. Haider, A., Haider, S. & Kang, I. REVIEW A comprehensive review summarizing the effect of electrospinning parameters and potential applications of nanofibers in biomedical and biotechnology. *Arab. J. Chem.* **11**, 1165–1188 (2018).

18. Bagbi, Y., Pandey, A. & Solanki, P. R. Chapter 10 - Electrospun Nanofibrous Filtration Membranes for Heavy Metals and Dye Removal. in *Nanoscale Materials in Water Purification* (eds. Thomas, S., Pasquini, D., Leu, S.-Y. & Gopakumar, D. A.) 275–288 (Elsevier, 2019). https://doi.org/10.1016/B978-0-12-813926-4.00015-X.

19. Liu, Z. *et al.* Electrospun Jets Number and Nanofiber Morphology Effected by Voltage Value: Numerical Simulation and Experimental Verification. *Nanoscale Res. Lett.* **14**, (2019).

20. Deitzel, J. M., Kleinmeyer, J., Harris, D. & Beck Tan, N. C. The effect of processing variables on the morphology of electrospun nanofibers and textiles. *Polymer (Guildf).* **42**, 261–272 (2001).

21. Abdulhussain, R., Adebisi, A., Conway, B. R. & Asare-Addo, K. Electrospun nanofibers: Exploring process parameters, polymer selection, and recent applications in pharmaceuticals and drug delivery. *J. Drug Deliv. Sci. Technol.* **90**, 105156 (2023).

22. Zargham, S., Bazgir, S., Tavakoli, A., Rashidi, A. S. & Damerchely, R. The Effect of Flow Rate on Morphology and Deposition Area of Electrospun Nylon 6 Nanofiber. **7**, 42–49 (2012).

23. Luo, C. J., Stoyanov, S. D., Stride, E., Pelan, E. & Edirisinghe, M. Electrospinning versus fibre production methods: from specifics to technological convergence. *Chem. Soc. Rev.* **41**, 4708–4735 (2012).

24. Shao, H., Fang, J., Wang, H. & Lin, T. Effect of electrospinning parameters and polymer concentrations on mechanical-to-electrical energy conversion of randomly-oriented electrospun poly(vinylidene fluoride) nanofiber mats. *RSC Adv.* **5**, 14345–14350 (2015).

25. Tarus, B., Fadel, N., Al-Oufy, A. & El-Messiry, M. Effect of polymer concentration on the morphology and mechanical characteristics of electrospun cellulose acetate and poly (vinyl chloride) nanofiber mats. *Alexandria Eng. J.* **55**, 2975–2984 (2016).
26. Meechaisue, C. & Supaphol, Æ. P. Electrospun cellulose acetate fibers : effect of solvent system on morphology and fiber diameter. 563–575 (2007) https://doi.org/10.1007/s10570-007-9113-4.
27. Park, B. K. & Um, I. C. Effect of molecular weight on electro-spinning performance of regenerated silk. *Int. J. Biol. Macromol.* **106**, 1166–1172 (2018).
28. Koski, A., Yim, K. & Shivkumar, S. Effect of molecular weight on fibrous PVA produced by electrospinning. *Mater. Lett.* **58**, 493–497 (2004).
29. Sun, B. *et al.* Advances in three-dimensional nanofibrous macrostructures via electrospinning. *Prog. Polym. Sci.* **39**, 862–890 (2014).
30. Topuz, F., Satilmis, B. & Uyar, T. Electrospinning of uniform nanofibers of Polymers of Intrinsic Microporosity (PIM-1): The influence of solution conductivity and relative humidity. *Polymer (Guildf).* **178**, 121610 (2019).
31. Angammana, C. J. & Jayaram, S. H. Analysis of the Effects of Solution Conductivity on Electrospinning Process and Fiber Morphology. *IEEE Trans. Ind. Appl.* **47**, 1109–1117 (2011).
32. Yang, G. Z., Li, H. P., Yang, J. H., Wan, J. & Yu, D. G. Influence of Working Temperature on The Formation of Electrospun Polymer Nanofibers. *Nanoscale Res. Lett.* **12**, (2017).
33. Van-Pham, D.-T. *et al.* No Title. *Green Process. Synth.* **9**, 488–495 (2020).
34. Wang, C. *et al.* Electrospinning of Polyacrylonitrile Solutions at Elevated Temperatures. *Macromolecules* **40**, 7973–7983 (2007).
35. Clerck, P. W. Æ. K. De. The effect of temperature and humidity on electrospinning. 1357–1362 (2009) https://doi.org/10.1007/s10853-008-3010-6.
36. Szewczyk, P. K. & Stachewicz, U. The impact of relative humidity on electrospun polymer fibers: From structural changes to fiber morphology. *Adv. Colloid Interface Sci.* **286**, 102315 (2020).
37. Raksa, A., Numpaisal, P. & Ruksakulpiwat, Y. The effect of humidity during electrospinning on morphology and mechanical properties of SF/PVA nanofibers. *Mater. Today Proc.* **47**, 3458–3461 (2021).
38. Kossyvaki, D. *et al.* Highly Porous Curcumin-Loaded Polymer Mats for Rapid Detection of Volatile Amines. *ACS Appl. Polym. Mater.* **4**, 4464–4475 (2022).
39. Katsogiannis, K. A. G., Vladisavljević, G. T. & Georgiadou, S. Porous electrospun polycaprolactone (PCL) fibres by phase separation. *Eur. Polym. J.* **69**, 284–295 (2015).
40. Huang, C. & Thomas, N. L. Fabricating porous poly(lactic acid) fibres via electrospinning. *Eur. Polym. J.* **99**, 464–476 (2018).
41. Farhaj, S., Conway, B. R. & Ghori, M. U. Nanofibres in Drug Delivery Applications. *Fibers* **11**, (2023).
42. Geng, Y. & Williams, G. R. Developing and scaling up captopril-loaded electrospun ethyl cellulose fibers for sustained-release floating drug delivery. *Int. J. Pharm.* **648**, 123557 (2023).
43. Wu, J. *et al.* Electrospinning of PLA Nanofibers: Recent Advances and Its Potential Application for Food Packaging. *J. Agric. Food Chem.* **70**, 8207–8221 (2022).
44. Rathore, P. & Schiffman, J. D. Beyond the Single-Nozzle: Coaxial Electrospinning Enables Innovative Nanofiber Chemistries, Geometries, and Applications. *ACS Appl. Mater. & Interfaces* **13**, 48–66 (2021).
45. Moghe, A. K. & Gupta, B. S. Co-axial Electrospinning for Nanofiber Structures: Preparation and Applications. *Polym. Rev.* **48**, 353–377 (2008).
46. Song, T. *et al.* Encapsulation of self-assembled FePt magnetic nanoparticles in PCL nanofibers by coaxial electrospinning. *Chem. Phys. Lett.* **415**, 317–322 (2005).

47. Zhao-Xia Huang Jia-Wei Wu, S.-C. W. J.-P. Q. & Srivatsan, T. S. The technique of electrospinning for manufacturing core-shell nanofibers. *Mater. Manuf. Process.* **33**, 202–219 (2018).

48. Li, D. & Xia, Y. Direct Fabrication of Composite and Ceramic Hollow Nanofibers by Electrospinning. (2004).

49. Aytac, Z. & Uyar, T. 13 - Applications of core-shell nanofibers: Drug and biomolecules release and gene therapy. in *Core-Shell Nanostructures for Drug Delivery and Theranostics* (eds. Focarete, M. L. & Tampieri, A.) 375–404 (Woodhead Publishing, 2018). https://doi.org/10.1016/B978-0-08-102198-9.00013-2.

50. Singh, R., Ahmed, F., Polley, P. & Giri, J. Fabrication and Characterization of Core–Shell Nanofibers Using a Next-Generation Airbrush for Biomedical Applications. *ACS Appl. Mater. & Interfaces* **10**, 41924–41934 (2018).

51. He, J. & Zhou, Y. Chapter 6 - Multineedle Electrospinning. in *Electrospinning: Nanofabrication and Applications* (eds. Ding, B., Wang, X. & Yu, J.) 201–218 (William Andrew Publishing, 2019). https://doi.org/10.1016/B978-0-323-51270-1.00006-6.

52. Tomaszewski, W. & Szadkowski, M. Investigation of electrospinning with the use of a multi-jet electrospinning head. *FIBRES & Text. East. Eur.* **13**, 22–26 (2005).

53. Theron, S. A., Yarin, A. L., Zussman, E. & Kroll, E. Multiple jets in electrospinning: experiment and modeling. *Polymer (Guildf).* **46**, 2889–2899 (2005).

54. Yang, Y. *et al.* A shield ring enhanced equilateral hexagon distributed multi-needle electrospinning spinneret. *IEEE Trans. Dielectr. Electr. Insul.* **17**, 1592–1601 (2010).

55. Xie, S. & Zeng, Y. Effects of Electric Field on Multineedle Electrospinning : Experiment and Simulation Study. (2012).

56. Kim, G., Cho, Y.-S. & Kim, W. D. Stability analysis for multi-jets electrospinning process modified with a cylindrical electrode. *Eur. Polym. J.* **42**, 2031–2038 (2006).

57. Kouhi, M., Mobasheri, M. & Valipouri, A. Chapter 7 - Needleless electrospinning. in *Electrospun and Nanofibrous Membranes* (eds. Kargari, A., Matsuura, T. & Shirazi, M. M. A.) 145–171 (Elsevier, 2023). https://doi.org/10.1016/B978-0-12-823032-9.00011-8.

58. Sasithorn, N., Martinová, L., Horáková, J. & Mongkholrattanasit, R. Fabrication of Silk Fibroin Nanofibres by Needleless Electrospinning. in *Electrospinning* (eds. Haider, S. & Haider, A.) (IntechOpen, Rijeka, 2016). https://doi.org/10.5772/65835.

59. Application, F. & Data, P. United States Patent (19). (1979).

60. Yarin, A. L. & Zussman, E. Upward needleless electrospinning of multiple nanofibers. *Polymer (Guildf).* **45**, 2977–2980 (2004).

61. Kotek, V., Martinova, L., Chaloupek, J., Examiner, P. & Lynn, M. (12) United States Patent. **2**, (2009).

62. Maleki, F., Razmi, H., Rashidi, M.-R., Yousefi, M. & Ghorbani, M. Recent advances in developing electrochemical (bio)sensing assays by applying natural polymer-based electrospun nanofibers: A comprehensive review. *Microchem. J.* **197**, 109799 (2024).

63. Liu, K. *et al.* Core-Shell Nanofibrous Materials with High Particulate Matter Removal Efficiencies and Thermally Triggered Flame Retardant Properties. *ACS Cent. Sci.* **4**, 894–898 (2018).

64. Li, T. *et al.* Anchoring CoFe2O4 Nanoparticles on N-Doped Carbon Nanofibers for High-Performance Oxygen Evolution Reaction. *Adv. Sci.* **4**, 1700226 (2017).

65. Ranjith, K. S., Celebioglu, A., Eren, H., Biyikli, N. & Uyar, T. Monodispersed, Highly Interactive Facet (111)-Oriented Pd Nanograins by ALD onto Free-Standing and Flexible Electrospun Polymeric Nanofibrous Webs for Catalytic Application. *Adv. Mater. Interfaces* **4**, 1700640 (2017).

66. Zhou, Z. *et al.* Graphene-beaded carbon nanofibers with incorporated Ni nanoparticles as efficient counter-electrode for dye-sensitized solar cells. *Nano Energy* **22**, 558–563 (2016).

67. Joly, D., Jung, J. W., Kim, I. D. & Demadrille, R. Electrospun materials for solar energy conversion: Innovations and trends. *J. Mater. Chem. C* **4**, 10173–10197 (2016).

68. Mwiiri, F. K. & Daniels, R. Chapter 3 - Electrospun nanofibers for biomedical applications. in *Delivery of Drugs* (ed. Shegokar, R.) 53–74 (Elsevier, 2020). https://doi.org/10.1016/B978-0-12-817776-1.00003-1.

69. Yan, B., Zhang, Y., Li, Z., Zhou, P. & Mao, Y. Electrospun nanofibrous membrane for biomedical application. *SN Appl. Sci.* (2022) https://doi.org/10.1007/s42452-022-05056-2.

70. Rasouli, R., Barhoum, A., Bechelany, M. & Dufresne, A. Nanofibers for Biomedical and Healthcare Applications. *Macromol. Biosci.* **19**, 1800256 (2019).

71. Online, V. A. *et al.* Electrospinning of recycled PET to generate tough mesomorphic fi bre membranes for smoke. 1632–1640 (2015) https://doi.org/10.1039/c4ta06191h.

72. Schinwald, A. *et al.* The Threshold Length for Fiber-Induced Acute Pleural Inflammation: Shedding Light on the Early Events in Asbestos-Induced Mesothelioma. *Toxicol. Sci.* **128**, 461–470 (2012).

Introduction

Electrospinning is a simple and highly versatile technique used for the production of ultrathin nanofibers. Electrospun nanofibers distinguished by their large surface area-to-volume ratio, high porosity, easy processability, and surface functionalization are excellent class of nanomaterials for an array of applications such as filtration membranes, drug delivery systems, tissue engineering scaffolds, wound dressings, protective textiles, and energy storage devices. Material selectivity and the ability to encapsulate various active agents are some other features which make electrospun nanofibers as an attractive choice for innovative applications.

Electrospun nanofibrous membranes have similarity in structure with our extracellular matrix, hence suitable for fabricating tissue engineering scaffolds to facilitate cell growth and regeneration. They are also used for the controlled and continuous release of therapeutic agents and an excellent alternative to traditional wound dressings. In environmental applications electrospun nanofibers acts as excellent filtration membranes for air and water filtration. The highly porous structure and large specific surface area effectively capture particulate matters and various pollutants present in air and water. In textile industry, electrospun nanofibers can be used to make functional fabrics with advanced features like superhydrophobic fabrics, flame retardant fabrics, antibacterial textiles with the same comfort and breathability of traditional cotton fabrics. Nanofibers show great potential in energy storage applications due to their unique structural properties. Electrospun nanofibers can be used to make electrodes and separators for batteries and supercapacitors. Their versatility allows the incorporation of various materials like carbon, metal oxides, and conductive polymers which further enhances their electrochemical performance. This chapter covers the detailed applications of electrospun nanofibers in biomedical, environmental, textiles, and energy storage devices.

© The Author(s), under exclusive license to Springer Nature Switzerland AG 2025   19
M. Mathew et al., *Electrospun Porous Nanofibers*, Synthesis Lectures on Emerging
Engineering Technologies, https://doi.org/10.1007/978-3-031-86106-2_2

Biomedical Applications

Electrospinning is an efficient and affordable technology for the development of nanofibers with distinctive characteristics such as large surface area-to-volume ratio, high porosity, and tunable surface properties. The small fiber diameter and high porosity of nanofibers makes them similar in structure to the native extracellular matrix, which improves the cell interactions. These characteristics make electrospun nanofibers suitable for various biomedical applications such as tissue engineering scaffolds, drug delivery, and wound dressings. The possibilities for surface functionalization and modification further enhance their applications in advanced biomedical fields. The polymers generally used for making nanofibers in biomedical applications include, polycaprolactone (PCL), polyvinyl alcohol (PVA), chitosan, collagen, cellulose acetate and many more, which are biodegradable and biocompatible [1].

Tissue Engineering Scaffolds

The goal of tissue engineering is to restore or replace damaged human tissues by combining cells with biomaterials and bioactive molecules. Scaffolds are three dimensional structures that can be made from a variety of biomaterials that can mimic the extracellular matrix (ECM) of the tissue to being engineered. An ideal scaffold for connective tissue engineering should possess several features to promote cellular biological processes. Some of the features other than to mimic ECM includes, high specific surface area, interconnected pores to facilitate blood vessel formation, and the ability for functionalization. The materials used as scaffolds for tissue engineering applications should be biodegradable, biocompatible, non-toxic, non-mutagenic and non-immunogenic. Biopolymers, both natural and synthetic are dominant materials in tissue engineering applications [2]. Several studies have revealed that the architecture of biopolymeric electrospun nanofibers best fits this application. Figure 2.1 represents typical human tissues with fibrous structure that can be regenerated with electrospun nanofibrous scaffolds [3].

Canciani et al. [4] fabricated polyblend nanofibers for the generation of gingival tissue. One of the main challenges faced by dental practitioners is the regeneration of oral soft tissue affected by periodontal disease was targeted using PCL nanofibers enriched with hyaluronic acid and vitamin E produced via electrospinning technique. Human fibroblast isolated from one gingival tissue fragment (HGF) was seeded onto nanofibers, preliminary results revealed that PCL-enriched nanofibers serve as a scaffold to effectively regenerate gingival soft tissues.

Drug Delivery Systems

Electrospun nanofibers have emerged as an excellent platform for the encapsulation and controlled release of therapeutic drug molecules. Conventional drug delivery systems such as tablets, capsules, and powders have certain drawbacks such as premature degradation, severe side effects of drugs, lack of target specificity, and discomfort during administration

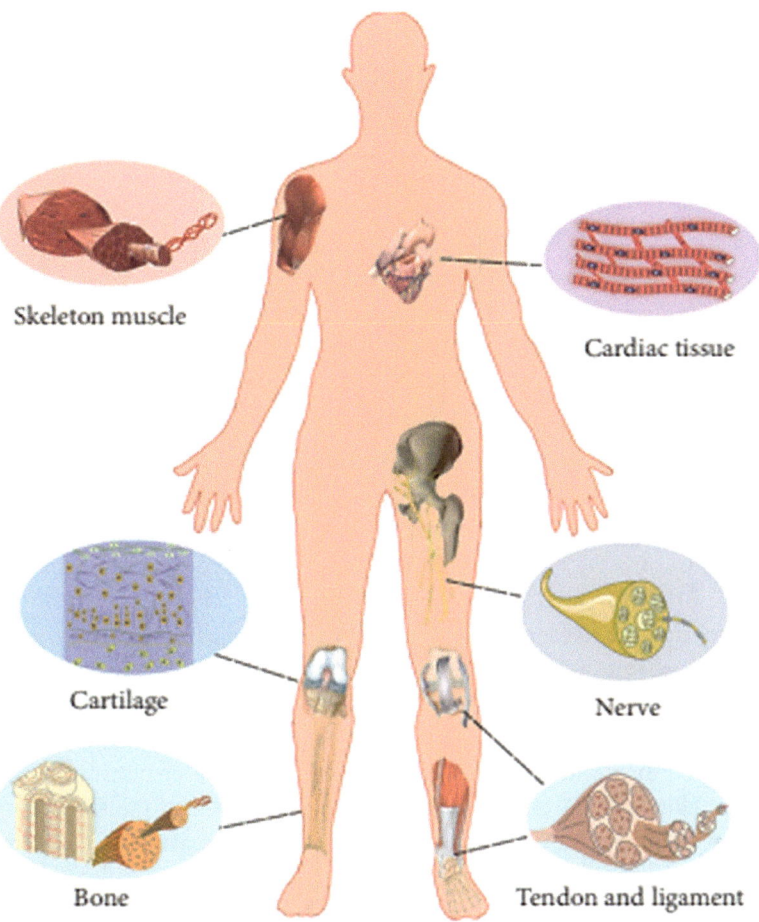

Skeleton muscle

Cardiac tissue

Cartilage

Nerve

Bone

Tendon and ligament

Fig. 2.1 Typical tissues with fibrous structures that can be regenerated using electrospun nanofibrous scaffolds [3]

[5]. To overcome these limitations, drugs can be encapsulated on to various nanocarriers like liposomes, polymeric micelles, nanoparticles, nanogels, and dendrimers. These nanocarriers can improve the solubility of drug, protect therapeutic agents from harsh physiological conditions, and provide controlled release characteristics. Compared to these nanocarriers, electrospun nanofibers have numerous distinctive features that makes them suitable as drug delivery systems. The highly porous structure of nanofibers provides high drug loading, high encapsulation efficiency, and reduced side effects.

Many efforts have been made to comprehend the drug release mechanism from nanofibers [6]. Fabrication methods, fiber morphology, polymer used, and drug-polymer interactions can affect the release kinetics. There are two main requirements for achieving

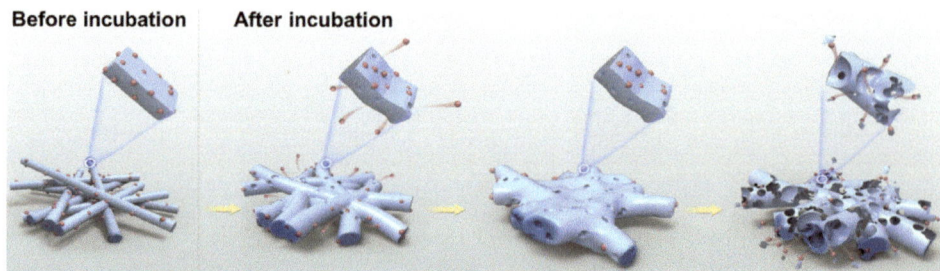

Before incubation   After incubation

Fig. 2.2 Different stages of sustained drug release mechanism from nanofibers. Reprinted with permission from Ref. [8]. Copyright Elsevier 2020

sustained release: (1) similarity in polarity between the drug and the polymer, (2) complete solubility of drug in the polymer solution [7]. Drug release from polymeric nanofibers is primarily a diffusion-driven process, that follows a Fickian model in biodegradable polymers and a non-Fickian in non-biodegradable systems. Wu et al. [8] described a three-step release mechanism for the release of ciprofloxacin hydrochloride drug from poly (lactic-co-glycolic acid) (PLGA)-based nanofibers. First step involves the diffusion of surface drug molecules into the release media, which causes the release medium to penetrate into the inter-fiber pores resulting in the swelling of nanofibers. The swelling provides the drug molecules more exposure to the release media, allowing further diffusion in accordance with the Fick's second law. In the second stage, the diffusion occurs slowly due to the fused fibers. Finally, the polymer degradation begins via enzymatic hydrolysis, further assisting the release of complete drug molecules from both the surface and bulk of nanofiber matrix. Figure 2.2 is schematic representation demonstrating the three stages of sustained drug release from nanofibers.

Wound Dressings

Wound dressings are normally used to protect the injured area from external environment, prevent bleeding, remove exudates, and activate clotting. Traditional materials such as cotton, bandage, and gauze are widely used for wound dressing applications. The drawbacks of traditional dressings are many: it may cause dehydration, prolonged exposure leads to formation of adhesions which delays the wound healing process and so on [9]. Electrospun nanofibrous membranes with high porosity, high encapsulation efficiency, and high drug loading capacity are suitable materials for wound dressings [10]. Their high specific surface area promotes gas exchange, absorption of wound exudates and the porous structure maintains moisture balance preventing excessive dryness and fluid accumulation. Further, electrospun nanofibers loaded with drug molecules or antibacterial agents prevent infection and accelerate the wound healing process. Both natural and synthetic biopolymers with good biocompatibility and biodegradability can be used as nanofiber wound dressings.

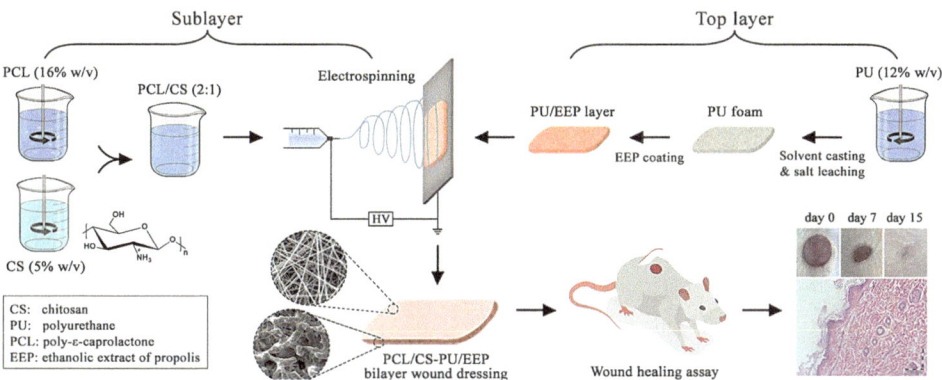

Fig. 2.3 PCL/CS-PU/EEP bilayer elctrospun membrane as wound dressing. Reprinted with permission from Ref. [11]. Copyright Elsevier 2022

Figure 2.3 represents the fabrication and application of a bilayer wound dressing comprising an electrospun polycaprolactone/chitosan (PCL/CS) fibrous mat as the base layer and ethanolic extract of propolis (EEP) coated polyurethane (PU) foam as the top layer [11]. PCL is a synthetic biopolymer having good biocompatibility, excellent mechanical flexibility, and processability. However, its slow biodegradation limits its applications. Hence, PCL was blended with chitosan, a natural polymer with good biocompatibility, biodegradability, hydrophilicity, and bacteriostatic properties. In order to overcome the poor mechanical strength by a single layered nanofibrous scaffold, a top layer with appropriate mechanical and antibacterial properties was added. So here PU with excellent mechanical properties and oxygen permeability coated propolis extract act as the top layer. Propolis is an antiseptic which prevents microbial infections. The in vitro and in vivo studies revealed that PCL/CS-PU/EEP bilayer wound dressing showed enhanced cell biocompatibility, antibacterial, and wound healing properties.

Environmental Applications

Environmental pollution is a global threat to both human health and economic development. Chemically toxic compounds and heavy metal ions are the primary sources causing most of the environmental pollutions [12]. The exposure to contaminations in air and water can cause severe health problems like allergy, cardiovascular disorders and infant mortality. Recently, a large number of pollutants including metal ions, pharmaceuticals, dyes, oil spills and pesticides have been detected in water which itself is a matter of great concern [13–15]. Air pollution sources include ozone, organic volatile compounds (ammonia, gasoline, formaldehyde), greenhouse gases (SO_x and NO_x) and airborne particulate matters ($PM_{2.5}$ and PM_{10}) [16].

Considering the serious impacts caused by these pollutants, tremendous efforts have been made by the research community to develop novel technology and devices for a clean environment. The most commonly used devices in daily life are in home active carbon-based water purifiers and portable air filters. However, these conventional devices have some shortcomings such as low adsorption efficiency, limited durability, and most of them are not reusable. With the advancement in nanoscience and nanotechnology, functional nanofibrous membranes have gained huge attention due to their excellent properties over conventional technologies [17]. Some of the advantages include their large surface area, small fiber diameter, highly porous structure, easy functionalization and cost effectiveness. Due to these features electrsopun nanofibers are prominent in environmental applications.

Air Filtration

Eletcrospun nanofiber-based air filters is found to be a simple and efficient method that allows the effective removal of air pollutants. Activated carbon and fiberglass are among the most commonly used materials in air filtration applications [18]. Activated carbon filters works by adsorption to trap the pollutants in air [19]. Potentially, membrane filtration can serve as a physical barrier that effectively removes various particulates, pollutants, volatile organic compounds (VOCs) and microorganisms depending upon the membrane pore size. Conventional micro-sized fiber membranes are not sufficient enough to remove smaller contaminants due to their large pore size [20]. At this point, electrospun nanofiber membranes are receiving more and more attraction due to their incredible filtration performance offered by their structural features [21]. Many polymers such as polyimide (PI), polyurethane (PU), polyacrylonitrile (PANI), polycaprolactone (PCL), polysulfone and other biopolymers have been successfully electrospinned to form nanofiber air filtration membranes [22–24]. The air filtration through electrospun nanofiber membrane occurs through five major mechanisms: (1) interception, (2) inertia, (3) diffusion, (4) gravity and (5) electrostatic. The interception effect occurs when particles near a fiber's surface are captured due to their size relative to the fiber's proximity. The inertia effect involves particles deviating from their airflow path and colliding with fibers due to their mass and velocity. Diffusion causes smaller particles to undergo Brownian motion, enhancing their likelihood of fiber contact and capture. Gravity can cause particles to settle onto fibers, though this is minimal for very small particles due to the short filtration time. Electrostatic effects occur when fibers and particles carry opposite charges, attracting particles to the fibers. Filtration efficiency depends on particle size: diffusion and electrostatic effects dominate for particles smaller than 0.3 µm, while interception and inertia are more effective for larger particles [25]. Multifunctional air filtration membranes are possible with the incorporation of antibacterial agents, various fillers and additives to the electrospinning solution. In a study conducted by Chueachot and co-workers [26] fabricated electrospun hybrid nanofibrous membranes of chitosan/silver nanoparticles/PVA/cellulose acetate with outstanding antibacterial activity. Also, the membrane showed high filtration efficiency of 99.78% ($PM_{2.5}$). Dual-spinneret electrospinning technique was used for the fabrication membrane. Transparent polyurethane (PU) air filter for the efficient capture of $PM_{2.5}$ was developed by Wen et al. [27]. They used a rotating bead spinneret for the large scale spinning of thermoplastic PU on a conductive mesh. This air filtration membrane has got high filtration efficiency and excellent ventilation, making it highly effective for indoor air purification.

Water Treatment

Electrospun nanofibers have been extensively utilized as filter for various water treatment applications like desalination, oil/water separation, surface water treatment, removal of toxic heavy metal ions and dyes [28]. Due to the precise control over pore sizes, nanofiber membranes provide enhanced filtration efficiency in waste water treatment plants. The membrane purification process relies on pressure driven-mechanism that acts as a selective barrier to separate impurities from usable water [29]. Based on the porosity of the electrospun membranes, membrane filtration can be classified into four; (1) ultrafiltration (pore size: 20 nm–0.1 μm, pressure range: 1–10 bars), (2) microfiltration (pore size: 0.1–5 μm, pressure: 0.1–3 bars), (3) reverse osmosis (pressure: 20–100 bars), (4) nanofiltration (pore size: 1–5 nm, pressure: 5–30 bars) [30]. Also, the interconnected pores create a large number of adsorption sites which enhances the adsorption efficiency of nanofiber membranes.

Oil/water separation is one of the critical issues faced by various industries. In a study conducted by Zhang et al. [31] they reviewed about various electrospun stretchable membranes for oil/water separation. The wettability of nanofibrous membrane is an important surface property in oil/water separation. Oil/water mixtures are typically separated using nanofibrous membranes that have an affinity for one phase while repelling the other. This separation requires membranes with selective wettability, such as hydrophilicity combined with oleophobicity or hydrophobicity paired with oleophilicity. A superhydrophobic polycaprolactone/beeswax (PCL-BW) electrospun membrane for efficient oil/water separation was successfully fabricated by Reshmi et al. [32]. The PCL-25BW membrane, containing 25 wt% beeswax demonstrated a superhydrophobic water contact angle of 153° and achieved 98% separation efficiency with gravity driven oil/water separation (Fig. 2.4). Heavy metals are another category of main pollutants in industrial wastewater. Major industrial activities responsible for these heavy metal ion contaminations are machinery manufacturing, electroplating, oil refining, mineral smelting and chemical processes. The major toxic heavy metals [33] include chromium (Cr), lead (Pb), cadmium (Cd), mercury (Hg), copper (Cu) and so on. The conventional methods used for the removal of heavy metals from wastewater include physical adsorption, ion exchange, electrolysis, chemical precipitation and membrane separation. Adsorption is the common method among all. An adsorbent with large number of adsorption sites, good adsorption capacity, easy separation and recovery are always preferred [34]. Activated carbon, graphene and zeolites are few examples for such adsorbents. At present, functionalized electrospun nanofibrous

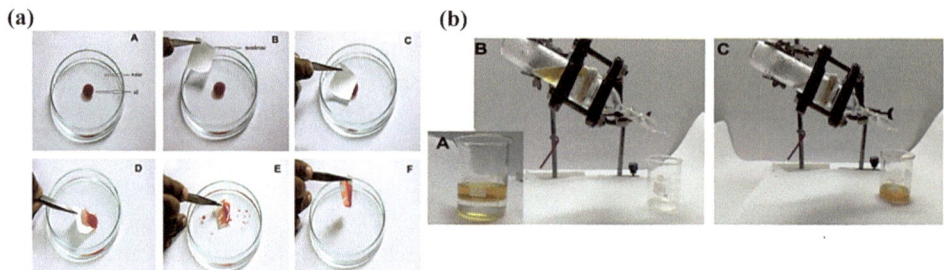

Fig. 2.4 **a** Selective absorption of oil from the water surface by PCL-25BW electrospun membrane; **b** gravity-assisted petrol/water separation setup using PCL-25BW membrane. Reproduced from Ref. [32] with permission from the Royal Society of Chemistry

membranes are extensively utilized for these applications due to their unique characteristics. Functional polymers containing amino, hydroxy, carboxyl, sulfonyl functional groups can be used for fabricating membranes having good adsorption affinity for heavy metals. Many natural polymers with poor spinnability have high adsorption affinity for heavy metals, therefore their adsorption performance can be enhanced by blending with synthetic polymers and organic/inorganic compounds. A novel electrospun nanofibrous membrane composed of chitosan (CS), polyvinylpyrrolidone (PVP), and polyvinyl alcohol (PVA) was developed by Wu et al. [35] to eliminate heavy metal ions and organic pollutants from water. Figure 2.5 is a schematic diagram showing the synthesis and application of CS/PVP/PVA nanofibrous membrane for heavy metal ions and organic dyes. Chitosan is a polycationic biopolymer rich in amino and hydroxyl functional groups that can form stable metal complexes by chelating with cationic metal ions and also, they can adsorb anionic metal ions through electrostatic interactions. However, due to their poor spinnability chitosan is often mixed with non-ionic polymers such as PVP. The large number of hydroxyl groups in PVP adsorbs heavy metal ions through hydrogen bonding and crosslinking. In this study, the adsorption performance of CS/PVP/PVA membranes were evaluated with Cu(II), Ni(II), Cd(II), Pb(II) ions, Malachite Green (MG), and Methylene Blue (MB) as target dyes. The membrane exhibited excellent adsorption capacities and excellent filtration efficiency with pure water permeate flux up to 4518.91 $Lm^{-2} h^{-1} bar^{-1}$.

Textile Applications

Electrospun nanofibers have many applications in textile industry including, textile functionality, antibacterial textiles, protective textiles, wound dressings and many more. Electrospun nanofibers due to their unique properties can be used to develop advanced textiles with enhanced performance and novel functionalities [36].

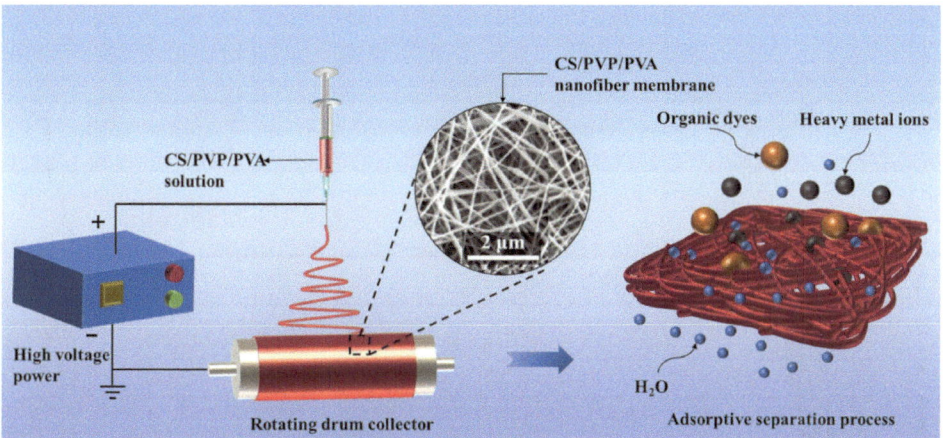

Fig. 2.5 Schematic diagram showing the synthesis and application of CS/PVP/PVA nanofibrous membrane for heavy metal ions and organic dyes. Reprinted with permission from Ref. [35]. Copyright Elsevier 2022

Protective Textiles

Electrospun nanofibers are widely used in protective textiles requiring high barrier performance. The ultrafine structure of nanofibers offers excellent breathability, filtration capability, while maintaining protective barrier against harmful chemicals. These features make electrospun nanofiber-based fabrics ideal for personal protective materials such as masks, gloves, and military clothing's. Additionally, the nanofibers can be functionalized with antimicrobial agents or it can be coated with nanoparticles to improve their protective capabilities, offering protection against chemical warfare and toxic chemicals. In a study conducted by Li et al. [37], they fabricated ceramic fibrous mat for protective clothing based electrospun SiO_2-TiO_2 fiber. TiO_2 was introduced into the silica fiber to control the thermal conductivity and optical properties of the non-woven fabric. The fabricated electrospun mat demonstrated exceptional flexibility, compressibility, and reduced thermal conductivity, making it ideal for use as an internal barrier layer in thermal protective clothing.

Functional Textiles

Functional textiles are specialized materials designed to provide additional properties beyond their primary function. Functional properties can be introduced at various stages of production, including the fiber, yarn, fabric, or garment stages. Electrospun nanofibers can be engineered to provide various functionalities including antibacterial, self-cleaning, UV protection, moisture wicking, and wrinkle resistance. Air permeability is a key factor in clothing comfort which is closely related to the porosity of electrospun nanofibers. The high porosity of nanofibers (90%) provides numerous channels for the passage of air

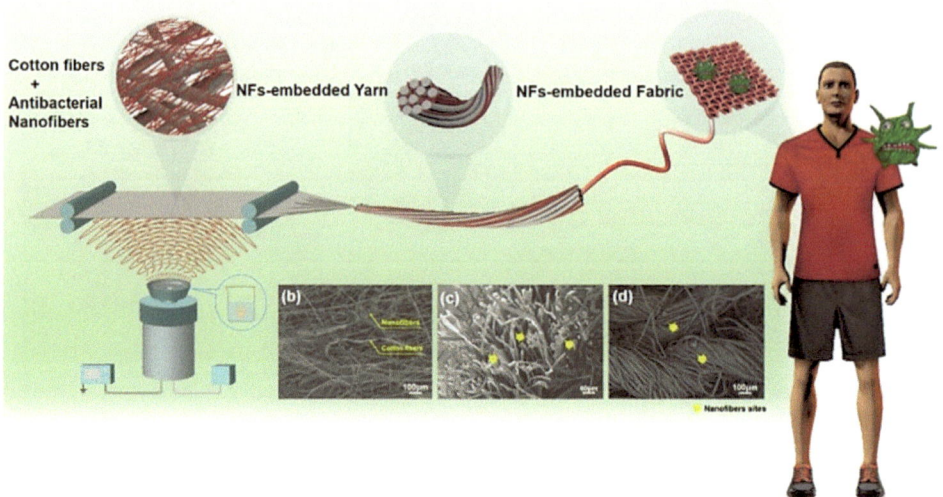

Fig. 2.6 Schematic representation on the fabrication of nanofiber-embedded textiles for sportswear applications. Reprinted with permission from Ref. [39]. Copyright Elsevier 2020

through the fabric. Electrospun nanofibers can be used as a coating in rain coats by optimizing the pore size that allows the selective permeability of water vapor, while blocking the rain droplets. Electrospun polyurethane (PU) mats have demonstrated superior water vapor and air permeability than PU and polytetrafluoroethylene (PTFE) coated fabrics [38].

Antibacterial textiles based on electrospun nanofibers play an important role in public health. Antimicrobial fabrics can prevent the growth of microorganisms, thus reducing the transmission of contagious diseases. One way to endow antibacterial properties with fabrics is to blend antibacterial fibers into textiles. In a work by Qui et al [39]. They introduced a sustainable and scalable approach for producing weavable antibacterial yarns by coating cotton fibers with antibacterial electrospun fibers. These coated fibers were then spun together into yarns (Fig. 2.6). The high content of cotton fibers ensured mechanical strength and weavability, while the antibacterial nanofibers impart antibacterial properties. The nanofiber-embedded textiles exhibited excellent antibacterial property, while maintaining the comfort of cotton fabrics. Sportswear based on this textile was fabricated, demonstrating enhanced durability of its antibacterial properties even after washing.

Smart Textiles

Smart textiles that respond to environmental changes such as temperature, humidity, light, electric field, and pH are receiving great interest these days. The importance of smart textiles lies in their ability to offer enhanced functionality and performance beyond traditional textiles. The exceptionally large specific surface area of electrospun fibers offers numerous interaction sites, enabling more effective interaction with the external environment and providing feedback. The feedback can be manifested through changes in physical, chemical properties, mechanical motions, shape change, or signal output. The development of electrospun mats in the realm of smart fibers has been greatly enhanced by stimuli-responsive polymers [40]. These polymers having high stimuli response when integrated with electrospinning, results in nanofibers that exhibit enhanced sensitivity due to their large surface area and high porosity.

Ma et al. reported the fabrication of thermosensitive drug loaded polylactic acid (PLA) core/ N-isopropylacrylamide (NIPAM) shell fibers via coaxial electrospinning. PLA fibers with surface porosity enhanced the surface area and drug loading capacity. PNIPAM is a thermo-sensitive polymer characterized by its unique lower critical solution temperature (LCST). It also exhibits reversible changes in volume and shape in response to small changes in temperature below and above the LCST. Thus, the temperature dependent state of PNIPAM shell significantly influenced the release of drug from PLA core.

Energy Applications

With rising global concerns over the exhaustion of fossil fuels and the increasing environmental challenges, the demand for efficient, sustainable energy storage solutions has increased drastically. Electrospun nanofibrous membranes have gained significant attention in energy storage applications due to their unique structural and functional properties. Nanofibrous membranes with high surface area-to-volume ratio, tunable porosity, and excellent flexibility are ideal for enhancing the performance of energy storage devices such as batteries, supercapacitors, and fuel cells. The higher surface area allows for better charge storage and interconnected porous structure makes ion transport faster, thus improving the overall efficiency. Moreover, the ability to incorporate wide range of materials such as conductive polymers, carbon fibers, and metal oxides further extends their functionality, making them favorable candidates for next-generation energy storage devices [41].

Battery Separators

Separators are one of the key components of battery devices. In a battery, the separator is a membrane positioned between the anode and cathode to prevent electrical short circuit, while facilitating the movement of ionic charge carriers during charging and discharging [42]. Based on the structure, separators can be classified into different categories like microporous membranes, gel polymer electrolytes, non-woven mats, and composite membranes. Considering the relevance of lithium-ion batteries as "green" energy storage devices, Costa and colleagues [43] reviewed the latest advancements in separator

Table 2.1 Electrolyte uptake (%) and ionic conductivity (mS/cm) of different types of separators

Type of separator	Electrolyte uptake (%)	Ionic conductivity (mS cm^{-1})
Porous PAN nanofiber	650	2.95
Non-porous PAN nanofiber	435	0.88
Celgard *PP*	218	0.86

membranes for lithium-ion batteries. Battery separators are typically made from porous membranes based on different polymers soaked in a liquid electrolytic solution. Some of the essential functions to be achieved by membrane-based separators include, (1) relatively thin separator with enough mechanical strength, (2) membranes with high porosity and low tortuosity to provide enough channels for ion mass transfer, (3) good affinity towards electrolyte, (4) should be electrochemically stable to avoid oxidation and reduction and (5) sufficient ionic conductivity to enhance battery capacity, cycle life, and overall cell resistance.

Electrospun nanofibrous membranes with high porosity, excellent air permeability, and enhanced electrolyte absorption are promising alternatives to traditional polyolefin separators (polypropylene, *PP*). Sabetzadeh et al. [44] prepared porous polyacrylonitrile (PAN) nanofiber separators for enhanced lithium-ion battery performance. Among polymer-based separators, PAN separators are attractive due to their high affinity towards electrolyte and excellent electrochemical stability. Porous structure was introduced via spinodal phase separation method. In this study, they compared the effect of membrane porosity on the battery performance using porous and non-porous PAN-nanofiber membrane with commercial Celgard *PP* separator. The porosity value for porous and non-porous PNA-nanofiber membranes were 83% and 72% respectively, while PP separators have only 45% porosity. The electrolyte uptake and ionic conductivity of each separator are tabulated in Table 2.1.

The charging-discharging capacities were measured by constructing $Li(Ni_{1/3}Co_{1/3}Mn_{1/3})O_2$/separator/Li metal coin half-cells. The half-cell with electrolyte-soaked PAN nanofiber membrane had higher charge discharge capacity (130 mAhg^{-1}) than other cells. These results revealed the enhanced performance and promising application of electrospun nanofiber-based membrane separators for new generation lithium-ion batteries.

Fuel Cell Electrodes

Recently, electrospun nanofibers have been widely used as cathode and anode in fuel cells due to their unique structural features. Their distinct porous structure enhances the electron and ion conductivity of nanofiber electrodes, thereby enhancing the electrochemical performance of fuel cells [45]. Fuel cells are devices that directly convert chemical energy into electrical energy. Based on the type of electrolytes, fuel cells can be categorized into solid oxide fuel cell (SOFC), phosphoric acid fuel cell (PAFC), alkaline fuel cell (AFC) and proton exchange membrane fuel cell (PEMFC).

Liu et al. [45] reviewed the potential applications of electrospun nanofibers in solid oxide fuel cells (SOFCs). SOFCs operate via oxygen-ion conduction, i.e.; reduction of oxygen molecules at the cathode to form oxygen ions and electrolyte carrying oxygen ions to the anode. At the anode, oxygen ions get chemically oxidized with a fuel (methane or hydrogen) to produce electrons to generate electricity. Nanofiber-based electrodes offer several advantages over traditional powder electrodes. These electrodes effectively reduce the activation barrier and expand the charge transport channels. In addition, the 3D-porous network creates a continuous charge transfer pathway, boosting the oxygen reduction reaction (ORR) activity, a feature that cannot be achieved with powder electrodes. Tang et al. fabricated cobalt-free nanofiber cathodes $La_2NiO_{4+\delta}$ (LNO) and $LaNi_{0.6}Fe_{0.4}O_{3-\delta}$ for proton conducting SOFCs. The cell with LNF nanofiber cathode demonstrated a reduced polarization resistance of 0.128 Ω cm^{-2} and a higher power output of 551 mWcm^{-2} at 700 °C.

Conclusion

Electrospinning is an efficient technique used to produce nanofibers with large surface area to volume ratio, high porosity, and tunable surface properties. Electrospun nanofibers produced with these unique structural features are well-suited for a broad spectrum of applications. In environmental applications, electrospun nanofibrous membranes are used as advanced filter for the efficient removal of pollutants from air and water. Nanofibers with interconnected porous structure mimic the extracellular matrix, this ability makes them suitable for applications in biomedical fields. In biomedical fields, nanofibers are widely used for the fabrication of tissue engineering scaffolds, wound healing patches, and for the controlled release of drug molecules. In textiles, nanofibers are increasingly applied to enhance the breathability, comfort, and functional properties. By incorporating bioactive agents, fabrics with multi-functional applications can be developed. Finally, in the energy sector the high surface area, and porous structure of nanofibers can enhance the performance of energy storage devices. Hence, nanofiber-based electrodes, separators, and electrolytes are widely used in batteries, supercapacitors, and fuel cells. Despite these advances, challenges for scalability, improved mechanical properties, and control over the fiber morphology still remains. Addressing these challenges is the key to unlocking the full potential of electrospun nanofibers.

References

1. Mwiiri, F. K. & Daniels, R. Electrospun nanofibers for biomedical applications. *Deliv. Drugs Vol. 2 Expect. Realities Multifunct. Drug Deliv. Syst.* 53–74 (2020) https://doi.org/10.1016/B978-0-12-817776-1.00003-1.

2. Asghari, F., Samiei, M., Adibkia, K., Akbarzadeh, A. & Davaran, S. Biodegradable and bio-compatible polymers for tissue engineering application: a review. *Artif. Cells, Nanomedicine Biotechnol.* **45**, 185–192 (2017).

3. Han, S. *et al.* 3D Electrospun Nanofiber-Based Scaffolds: From Preparations and Properties to Tissue Regeneration Applications. *Stem Cells Int.* **2021**, 1–22 (2021).

4. Canciani, E. *et al.* Polyblend Nanofibers to Regenerate Gingival Tissue: A Preliminary In Vitro Study. *Front. Mater.* **8**, 1–11 (2021).

5. Torres-Martinez, E. J., Cornejo Bravo, J. M., Serrano Medina, A., Pérez González, G. L. & Villarreal Gómez, L. J. A Summary of Electrospun Nanofibers as Drug Delivery System: Drugs Loaded and Biopolymers Used as Matrices. *Curr. Drug Deliv.* **15**, 1360–1374 (2018).

6. Luraghi, A., Peri, F. & Moroni, L. Electrospinning for drug delivery applications: A review. *J. Control. Release* **334**, 463–484 (2021).

7. Zeng, J. *et al.* Influence of the drug compatibility with polymer solution on the release kinetics of electrospun fiber formulation. *J. Control. Release* **105**, 43–51 (2005).

8. Wu, J. *et al.* Mechanism of a long-term controlled drug release system based on simple blended electrospun fibers. *J. Control. Release* **320**, 337–346 (2020).

9. Peng, W. *et al.* Recent progress of collagen, chitosan, alginate and other hydrogels in skin repair and wound dressing applications. *Int. J. Biol. Macromol.* **208**, 400–408 (2022).

10. Zhang, X. *et al.* Advances in wound dressing based on electrospinning nanofibers. *J. Appl. Polym. Sci.* **141**, e54746 (2024).

11. Shie Karizmeh, M., Poursamar, S. A., Kefayat, A., Farahbakhsh, Z. & Rafienia, M. An in vitro and in vivo study of PCL/chitosan electrospun mat on polyurethane/propolis foam as a bilayer wound dressing. *Biomater. Adv.* **135**, 112667 (2022).

12. Sobhanie, E., Roshani, A. & Hosseini, M. Chapter 16 - Microfluidic systems with amperometric and voltammetric detection and paper-based sensors and biosensors. in *Carbon Nanomaterials-Based Sensors* (eds. Manjunatha, J. G. & Hussain, C. M.) 275–287 (Elsevier, 2022). https://doi.org/10.1016/B978-0-323-91174-0.00023-8.

13. Ahmad, A. *et al.* Recent trends and challenges with the synthesis of membranes: Industrial opportunities towards environmental remediation. *Chemosphere* **306**, 135634 (2022).

14. Tkaczyk, A., Mitrowska, K. & Posyniak, A. Synthetic organic dyes as contaminants of the aquatic environment and their implications for ecosystems: A review. *Sci. Total Environ.* **717**, 137222 (2020).

15. Zaboon, S. *et al.* Removal of monoethylene glycol from wastewater by using Zr-metal organic frameworks. *J. Colloid Interface Sci.* **523**, 75–85 (2018).

16. Robert, B. & Nallathambi, G. A concise review on electrospun nanofibres/nanonets for filtration of gaseous and solid constituents (PM2.5) from polluted air. *Colloid Interface Sci. Commun.* **37**, 100275 (2020).

17. Saleem, H., Trabzon, L., Kilic, A. & Zaidi, S. J. Recent advances in nanofibrous membranes: Production and applications in water treatment and desalination. *Desalination* **478**, 114178 (2020).

18. Sundarrajan, S., Tan, K. L., Lim, S. H. & Ramakrishna, S. Electrospun Nanofibers for Air Filtration Applications. *Procedia Eng.* **75**, 159–163 (2014).

19. Roegiers, J. & Denys, S. Development of a novel type activated carbon fiber filter for indoor air purification. *Chem. Eng. J.* **417**, 128109 (2021).

20. Wang, N. *et al.* Superamphiphobic nanofibrous membranes for effective filtration of fine particles. *J. Colloid Interface Sci.* **428**, 41–48 (2014).

21. Zhu, M., Xiong, R. & Huang, C. Bio-based and photocrosslinked electrospun antibacterial nanofibrous membranes for air filtration. *Carbohydr. Polym.* **205**, 55–62 (2019).

22. Yu, Z. *et al.* Efficient air filtration through advanced electrospinning techniques in nanofibrous Materials: A review. *Sep. Purif. Technol.* **349**, 127773 (2024).

23. Nicosia, A. *et al.* Air filtration and antimicrobial capabilities of electrospun PLA/PHB containing ionic liquid. *Sep. Purif. Technol.* **154**, 154–160 (2015).

24. Wan, H. *et al.* Hierarchically structured polysulfone/titania fibrous membranes with enhanced air filtration performance. *J. Colloid Interface Sci.* **417**, 18–26 (2014).

25. Liu, H., Zhu, Y., Zhang, C., Zhou, Y. & Yu, D.-G. Electrospun nanofiber as building blocks for high-performance air filter: A review. *Nano Today* **55**, 102161 (2024).

26. Chueachot, R., Promarak, V. & Saengsuwan, S. Enhancing antibacterial activity and air filtration performance in electrospun hybrid air filters of chitosan (CS)/AgNPs/PVA/cellulose acetate: Effect of CS/AgNPs ratio. *Sep. Purif. Technol.* **338**, 126515 (2024).

27. Liang, W. *et al.* Transparent Polyurethane Nanofiber Air Filter for High-Efficiency PM2.5 Capture. *Nanoscale Res. Lett.* **14**, (2019).

28. Uddin, Z. *et al.* Recent trends in water purification using electrospun nanofibrous membranes. *Int. J. Environ. Sci. Technol.* **19**, 9149–9176 (2022).

29. Qu, X., Alvarez, P. J. J. & Li, Q. Applications of nanotechnology in water and wastewater treatment. *Water Res.* **47**, 3931–3946 (2013).

30. Ma, H., Burger, C., Hsiao, B. S. & Chu, B. Ultra-fine cellulose nanofibers: new nano-scale materials for water purification. *J. Mater. Chem.* **21**, 7507–7510 (2011).

31. Zhang, J. *et al.* Electrospun flexible nanofibrous membranes for oil/water separation. *J. Mater. Chem. A* **7**, 20075–20102 (2019).

32. C. R., R., Sundaran, S. P., A., J. & Athiyanathil, S. Fabrication of superhydrophobic polycaprolactone/beeswax electrospun membranes for high-efficiency oil/water separation. *RSC Adv.* **7**, 2092–2102 (2017).

33. Zhu, F., Zheng, Y.-M., Zhang, B.-G. & Dai, Y.-R. A critical review on the electrospun nanofibrous membranes for the adsorption of heavy metals in water treatment. *J. Hazard. Mater.* **401**, 123608 (2021).

34. Chen, T. *et al.* Adsorption of cadmium by biochar derived from municipal sewage sludge: Impact factors and adsorption mechanism. *Chemosphere* **134**, 286–293 (2015).

35. Wu, S., Li, K., Shi, W. & Cai, J. Preparation and performance evaluation of chitosan/polyvinylpyrrolidone/polyvinyl alcohol electrospun nanofiber membrane for heavy metal ions and organic pollutants removal. *Int. J. Biol. Macromol.* **210**, 76–84 (2022).

36. Mathew, M. & Rajan Paul, R. Nanofibers and Nanotechnology in Porous Textiles. in 237–253 (2024). https://doi.org/10.1201/9781003414469-13.

37. Li, J. *et al.* Electrospun silica-based fiber for developing multifunctional protective clothing. *Mater. Sci. Eng. B* **289**, 116259 (2023).

38. Wang, L. Functional Nanofibre: Enabling Material for the Next Generation Smart Textiles. *J. Fiber Bioeng. Informatics* **1**, 81–92 (2008).

39. Qiu, Q. *et al.* Functional nanofibers embedded into textiles for durable antibacterial properties. *Chem. Eng. J.* **384**, 123241 (2020).

40. Wei, M., Gao, Y., Li, X. & Serpe, M. J. Stimuli-responsive polymers and their applications. *Polym. Chem.* **8**, 127–143 (2017).

41. Sun, G., Sun, L., Xie, H. & Liu, J. Electrospinning of nanofibers for energy applications. *Nanomaterials* **6**, (2016).

42. Luiso, S. & Fedkiw, P. Lithium-ion battery separators: Recent developments and state of art. *Curr. Opin. Electrochem.* **20**, 99–107 (2020).

43. Costa, C. M., Lee, Y.-H., Kim, J.-H., Lee, S.-Y. & Lanceros-Méndez, S. Recent advances on separator membranes for lithium-ion battery applications: From porous membranes to solid electrolytes. *Energy Storage Mater.* **22**, 346–375 (2019).

44. Sabetzadeh, N., Gharehaghaji, A. A. & Javanbakht, M. Porous PAN micro/nanofiber separators for enhanced lithium-ion battery performance. *Solid State Ionics* **325**, 251–257 (2018).

45. Liu, Z., Gu, Y. & Bi, L. Applications of electrospun nanofibers in solid oxide fuel cells – A review. *J. Alloys Compd.* **937**, 168288 (2023).

Eletcrospun Nanofiber Membranes

Introduction

Electrospun nanofiber membranes are innovative materials produced through electrospinning process, a technique that generates ultrathin fibers from polymer solutions under the influence of electrostatic forces. These membranes are characterized by their distinctive features such as high specific surface area, high porosity, flexibility, and tunable mechanical properties. The structure and orientation of electrospun membranes can be controlled by modifying the solution parameters and electrospinning setup. Owing to their remarkable characteristics, electrospun nanofiber membranes have broad spectrum of applications, including biomedical implants, drug delivery, filtration systems, food packaging, biosensors, and energy storage devices. This chapter examines the structure, properties, and some selected applications of electrospun nanofiber membranes.

Structure and Properties of Nanofiber Membranes

Electrospun nanofiber membranes (ENMs) formed through electrospinning process exhibit many unique properties and structure. This includes large surface area-to-volume ratio, elevated porosity, adjustable pore size, high aspect ratio, tunable pore size, good thermal and mechanical stability with diverse fiber morphologies [1]. The superior mechanical properties of electrospun nanofibers compared to their bulk material is primarily due to the reduced material defects and higher molecular orientation [2]. Controlling fiber structure and morphology is very essential for ensuring consistent and reproducible performance. The morphology of these fibers can be controlled by adjusting both solution parameters (molecular weight, concentration, surface tension, and viscosity) and electrospinning parameters (applied voltage, flow rate, and needle tip-to-collector distance) [3,

© The Author(s), under exclusive license to Springer Nature Switzerland AG 2025
M. Mathew et al., *Electrospun Porous Nanofibers*, Synthesis Lectures on Emerging Engineering Technologies, https://doi.org/10.1007/978-3-031-86106-2_3

4]. Also, electrospinning allows the use of a wide range materials including both natural and synthetic polymers, as well as diverse solvents, both water and organic solvents. Electrospun nanofibers with highly interconnected porous structure enhances the functional properties by providing more active sites for improved interactions with biological or chemical agents, thus making them suitable for many applications.

Porosity and Pore Size Distribution

Porosity and pore size distribution of electrospun nanofibers are dependent on both solution parameters (concentration, viscosity) and electrospinning conditions (applied voltage, flow rate, working distance) [5]. Often these properties of are less investigated compared to their chemical and biochemical properties, yet these are very important for the intended application. Certain applications require high porosity, while some other requires low. Hence, it is essential to optimize the porosity during electrospinning process Most physical parameters such as hydrophobicity, air/water vapor permeability, water uptake, and cell adhesion properties are dependent on the porosity and pore size distribution [6].

Porosity refers to the volume of voids within a given volume of nanofiber membrane. The pores can exist between the interconnected fibrous structure of nanofibers, which is the inherent porous structure of nanofibers [7]. In addition to this, pores can be induced by various methods like phase separation, collecting nanofibers in liquid nitrogen bath, and polymer blending followed by the selective removal of one of the polymers [8]. Currently available methods for measuring porosity includes density method, image method, and mercury intrusion method. In density method, the bulk density or the density of a given material is compared with its theoretical density (density of raw material) [9]. Scanning electron microscopy (SEM) and transmission electron microscopy (TEM) are two dynamic tools that helps in the precise determination of pore size distribution and porosity [10]. SEM images can reveal the surface morphology of nanofibers, providing surface pore size distribution and overall porosity. While, TEM can capture images that reveal the internal porous structure of thin nanofiber membranes (Fig. 3.1).

In mercury intrusion porosimetry, mercury is forced into the pores of a porous structure at a high pressure. The pore size decreases as the pressure increases. By analysing the relationship between pressure and volume, mathematical models can be used to determine the pore size distribution [12].

Mechanical Properties

Electrospun nanofiber membranes possess outstanding mechanical properties due to their unique nano-structural features such as large surface area-to-volume ratio, excellent chemical and thermal stability, good flexibility, high porosity, and tunable surface properties

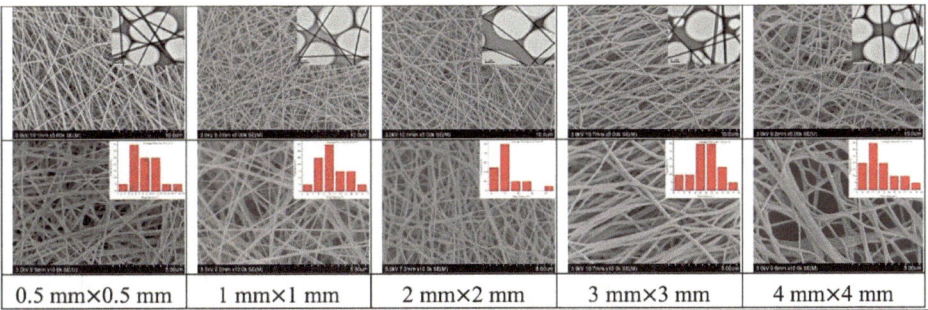

| 0.5 mm×0.5 mm | 1 mm×1 mm | 2 mm×2 mm | 3 mm×3 mm | 4 mm×4 mm |

Fig. 3.1 SEM, and TEM, images of electrospun fibers along with their corresponding pore size distribution. Reprinted with permission from Ref. [11]. Copyright Elsevier 2019

[13]. These properties make them ideal candidate for diverse applications such as filtration, biomedical implants, food packaging, sensors, and so on. Mechanical properties of electrospun nanofiber membranes are an important parameter that determines the suitability of their application [2]. Critical mechanical properties consist of tensile strength, elongation at break, elastic modulus, shear strength, compression strength, and toughness. All these properties depend on the polymer type, fiber morphology, alignment, additives and fillers used. Literature often reports on the mechanical behavior of electrospun nanofiber membrane rather than individual fibers, since measuring their properties is quite challenging. Generally, randomly oriented fibers are obtained in normal electrospinning process. However, aligned fibers usually exhibit advantageous properties such as good mechanical strength over random fibers, because the fibers are oriented in the same direction which increases their tensile modulus and strength. Han et al. provided a comprehensive review on electrospun nanofibers with aligned structures [14]. In particular, aligned electrospun nanofibers find applications in matrix reinforcement and in the development of energy storage devices. Among the various strategies to produce aligned fiber using electrospinning, two main strategies are the use of a stationary collector and the utilization of a rotating collector. Aligned polyethylene oxide (PEO) nanofibers were fabricated by Huang et al. through the modification of a conventional collector [15]. They used an additional frame to control the electric field distribution for guiding the distribution pathway of electrospun fibers. Using a rotating collector, Matthews et al. successfully fabricated aligned collagen fibers [16]. The mechanical performance of electrospun nanofibers is also influenced by other factors like fiber diameter, type of polymer, its concentration, and environmental conditions during electrospinning process.

Electrospun Nanofiber Membranes with Different Structures

Electrospinning is considered as a versatile technique that allows for the fabrication of fibers with different morphologies and structures such as core–shell, porous, and hollow nanofibers [17]. Conventional ENMs suffer from lower strength and difficulty in functionalization, which limits their wide range of applications. This limitation can be addressed through the fabrication of fibers with different morphologies and structures by the precise control of fabrication steps and by using special spinnerets. Fiber membranes with core shell, porous, and hollow structure significantly enhances the fiber properties such as surface area, mechanical strength, and provide additional functionalities [18]. Also, these structures open the way for innovative applications in drug delivery, tissue regeneration, catalysis, filtration, and energy storage devices.

Core–Shell Fibers

Core–shell structured nanofiber membranes are generally produced using four main methods: (1) coaxial electrospinning, (2) template deposition, (3) electrospinning of immiscible polymer blends and (4) emulsion electrospinning [19]. In template-based synthesis, the shell material is deposited over a fiber template using techniques such as chemical vapor deposition, plasma, or sol gel coating methods, which is a complex procedure [20]. Emulsion electrospinning is a novel technique used for fabricating core–shell fibers from water in oil (W/O) or oil in water (O/W) emulsions [21]. The concept of this technique relies on the solution using for electrospinning. In an emulsion, where two immiscible liquids are combined to a single system using emulsifier one liquid is dispersed as droplets within other liquid. In such a system, a drug that is soluble in an organic solvent can be incorporated into a hydrophilic medium, and vice versa [22]. Figure 3.2 is a schematic illustration of emulsion electrospinning. In this process, two immiscible polymers are emulsified prior to electrospinning. On applying a high electric voltage, the solvent near the surface of emulsion droplets evaporates quickly which causes a rapid increase in viscosity at the outer layer. Subsequently, the polymer jet undergoes inward movement, stretching, and elongation leading to the formation of core–shell fibers with dispersed droplets [23].

Coaxial electrospinning is the most commonly used technique among these four. This technique involves the use of a coaxial spinneret to feed two different polymer solutions to electrospun both solutions simultaneously. The outer and inner spinnerets carrying polymer solution is connected to a high voltage power supply, core and shell fibers are formed through solvent evaporation and stretching as in the case of conventional electrospinning process [24]. Figure 3.3 represents a general coaxial electrospinning setup. One of the greatest advantages of coaxial electrospinning is that it allows the production of dual-phase fibers, where both the core and the shell fibers retain their individual characteristic properties. Thus, the finally obtained structure would have a unique blend

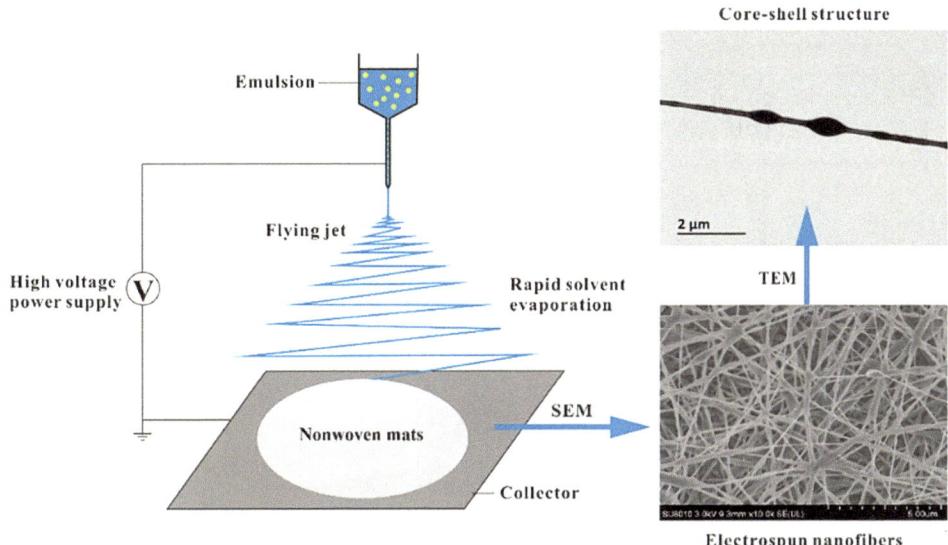

Fig. 3.2 Schematic illustration of emulsion electrospinning. Reprinted with permission from Ref. [21]. Copyright Elsevier 2018

of properties from both the materials. Compared to other methods for core–shell fiber production, coaxial electrospinning has many other advantages-single set process, highly efficient method that does not require any complex working conditions like vacuum, high temperature, or plasma treatments. It enables the combination of electrospinnable materials with those materials that are not spinnable, thus broadening the range of materials that can be made into coaxial nanofibers with hybrid properties and unique functions [25]. To achieve the successful fabrication of core–shell fibers, selection of solvents for the core and shell fibers is an important factor. Generally, either completely miscible or immiscible solutions are taken to minimize the solute–solvent interaction. The use of two immiscible solutions minimizes the interdiffusion between layers with good separation between core and shell fibers. Similarly, two miscible solutions can also generate distinct core–shell structure if the travelling time of polymer jet is shorter than the solution diffusion time constant [26]. Some other conditions for obtaining uniform core-sheath fibers include, the use of shell solution with high viscosity, conductivity, and flow rate than core solution, minimum surface tension at the core and shell interface, and shell solvent with lower volatility for stable Taylor cone formation.

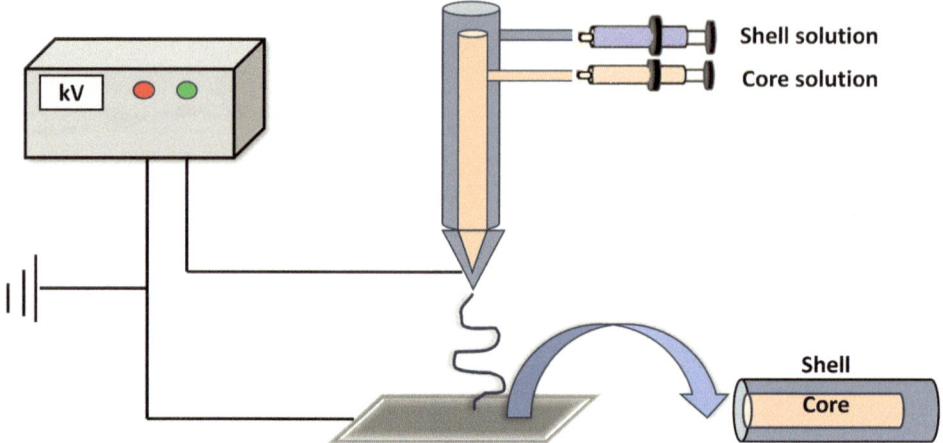

Fig. 3.3 Coaxial electrospinning setup

Porous Nanofibers

Electrospun porous nanofibers have gained a huge interest in various fields due to their exceptionally high surface area, tunable porous structure, abundant active sites. Easy functionalization, and light weight which can further enhance the performance of materials [27]. Different from the inherent pores between the fibers, porous nanofibers refer to the presence of voids or pores on the surface or within the fiber structure as shown in Fig. 3.4.

Depending on the pore size, porous fibers can be categorized into three: micropores (< 2 nm), mesopores (2–50 nm), and macropores (> 50 nm). There are several strategies for constructing porous structures in electrospun nanofibers. Some of the methods

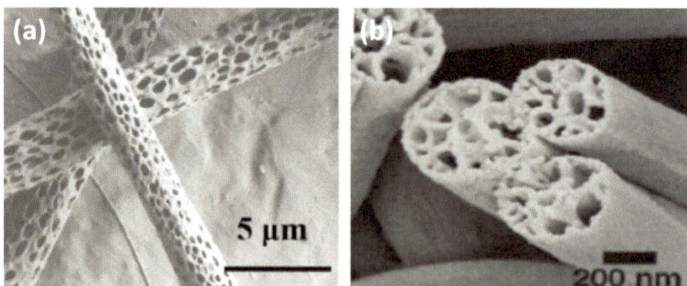

Fig. 3.4 **a** SEM image of porous PLA nanofiber (surface porous structure). Reprinted with permission from Ref. [28]. Copyright Elsevier 2021. **b** Cross-sectional SEM image of TiO$_2$ nanofibers (internal porous structure). Reproduced with permission from Ref. [29]. Copyright Springer Nature 2022

include phase separation mechanism, electrospinning of polymer blends followed by the sacrificial removal of one of the polymer, using liquid nitrogen bath for collecting fibers, and using salts or additives capable of generating porous structures [30]. In a study performed by Huang et al. [31] highly porous PLA nanofibers were fabricated via phase separation mechanism. Phase separation techniques are generally divided into thermally induced phase separation (TIPS), non-solvent induced phase separation (NIPS), and vapor induced phase separation (VIPS). In TIPS, the temperature gradient serves as the driving force for pore formation. Here, a highly volatile solvent is used for electrospinning the polymer solution followed by the rapid cooling of fibers by immersing in a liquid nitrogen bath, which completely removes the solvent to create pores. In NIPS, the phase separation is induced by the introduction of a highly volatile non-solvent to the electrospinning solution. The evaporation of the non-solvent generates pores on the fibers. In NIPS, porosity can be generated by electrospinning the polymer using a water miscible-low volatile solvent in a humid environment. During electrospinning, water molecules present in the environment get attracted towards the fiber and causes phase separation, since water act as a non-solvent for the polymer. This is followed by the complete removal of both solvent and water molecules. Figure 3.5 represents the SEM images of highly porous PLA fibers fabricated via vapor induced phase separation mechanism at humid environment using dimethyl sulfoxide (DMSO, boiling point $= 189°$) as non-solvent.

Nagamine et al. [32] fabricated porous carbon nanofibers by electrospinning an aqueous solution of polyvinyl alcohol (PVA), diammonium hydrogen phosphate (DAP), and sodium dodecyl sulphate (SDS) (Fig. 3.6). The PVA precursor solution was converted into carbon nanofibers via stabilization and carbonization processes. Here DAP was used as the stabilizing agent prior to carbonization. SDS was added to reduce the surface tension and to improve the spinnability. The thermal decomposition of DAP and SDS during stabilization and carbonization generates water soluble salt particles over the nanofibers. These particles act as temporary templates for the macropores, which can be easily removed with water.

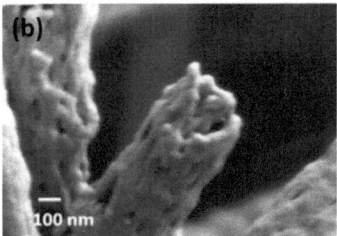

Fig. 3.5 SEM images of highly porous PLA fibers **a** surface porosity, **b** internal porosity. Reprinted with permission from Ref. [31]. Copyright Elsevier 2018

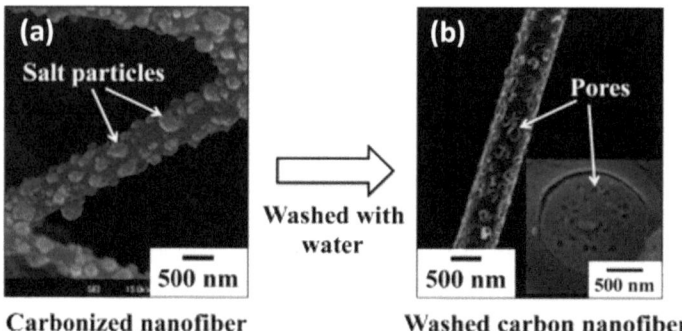

Fig. 3.6 SEM images of **a** surface of carbon nanofibers before washing, (b) surface of washed carbon nanofibers. Reprinted with permission from Ref. [32]. Copyright Elsevier 2016

Hollow Nanofibers

Hollow nanofibers are tubular structures with empty cores widely used in drug delivery, energy storage, catalysis, and biosensors. Hollow nanofibers can be generated either via chemical vapor deposition (CVD) or direct coaxial electrospinning [33 34]. In CVD, the polymer is first electrospun into fibers which then act as the template. Then over this template other materials such as polymer or metal is coated by CVD to form the outer shell. Finally, the electrospun template material is removed by using a suitable solvent or by calcination, leaving behind a hollow structure. Similarly, in co-axial electrospinning the core polymer is selectively removed at the end of the process. Kumar et al. [35] fabricated a nano catalyst based on hollow TiO_2 nanofibers embedded with gold nanoparticles (AuNPs) via co-axial electrospinning. The core forming solution involves a mixture of poly (4-vinyl pyridine) and AuNPs, whereas a mixture of poly (vinyl pyrrolidone) and titania sol constitute the shell solution. The core–shell fibers were fabricated using co-axial electrospinning technique, followed by the stepwise calcination from room temperature to 300 to 600 °C. This heat treatment removes poly (4-vinyl pyridine) components from the core and AuNPs are successfully entrapped in the inner walls of porous titania shell. Figure 3.7 represents the SEM images of Au@TiO_2 nanofibers before and after calcination.

Membrane Applications

Electrospun nanofibrous membranes can be used for many applications [36]. The fine fibers within the membrane structure provides high surface area, tunable porosity, and exceptional mechanical strength. Some common applications include filtration (air and water filtration), tissue engineering scaffolds (highly porous electrospun fibers mimic

Fig. 3.7 SEM images of Au@TiO$_2$ nanofibers **a** before calcination and **b** after calcination. Reproduced from Ref. [35] with permission from the Royal Society of Chemistry

extracellular matrix), drug delivery (allows controlled and targeted delivery of drug molecules), and sensors (excellent sensitivity and selectivity towards biomolecules). In this section we will discuss about a very few applications of electrospun nanofiber membranes.

Biomedical Implants

Electrospun nanofibers with highly porous structure that can mimic the extracellular matrix are excellent materials for biomedical implants. Their large surface area-to-volume ratio, flexibility, and high porosity can enhance the cell adhesion, proliferation and differentiation [37]. Among various biomaterials, eletcrospun nanofibers are ideal materials for replacing and reconstructing damaged tissues and organs bone, cartilage, skin, heart and blood vessels. One of the greatest advantages of electrospun nanofibers is the possibility to control their structure and morphology by optimizing both solution and electrospinning parameters. Thus, via surface functionalization their compatibility and performance when used as medical implants can be improved. Another important factor is that, the material should be biocompatible, biodegradable, and easily processable [38].

In a study conducted by Wu et al. [39], they fabricated biomineralized tetramethylpyrazine (TMP) loaded PCL/gelatin nanofibrous membrane for bone angiogenesis and repair. Polycaprolactone (PCL) is a commonly used polymer for biomedical applications, however its hydrophobicity limits their application. In order to overcome this issue, here PCL was blended with hydrophilic gelatin to improve its hydrophilicity and biocompatibility. Conventional electrospun fibrous membranes have smooth surface, which reduces their ability to support biomineralization. Also, the smooth surface hinders cell adhesion and proliferation which is essential for tissue regeneration. To enhance the surface roughness bubble-shaped nanofibers were fabricated by mixing with paraffin spheres.

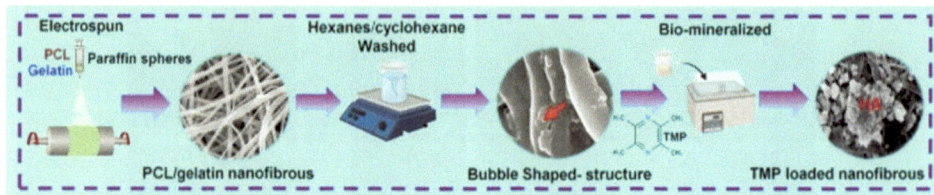

Fig. 3.8 Fabrication of bubble-shaped TMP loaded PCL/gelatin nanofibrous membrane for bone regeneration [39]

TMP drug which is widely used in the treatment of cardiovascular diseases was loaded on to the surface of membrane to promote vascularization and bone repair. In vitro analysis revealed that, PCL/gelatin fibers with bubble-shaped structure increased the surface roughness which promoted the mineralization. Also, the drug loaded membrane improved the cell attachment, proliferation and differentiation. When tested in vivo, the membrane further encouraged vascularization and facilitated bone formation in rat cranium defect models. Figure 3.8 is a schematic representation on the fabrication of bubble-shaped TMP loaded PCL/gelatin nanofibrous membranes.

Food Packaging Membranes

Food packaging is considered as a crucial step in food processing as it plays multiple functions. The principal roles of food packaging are to protect the food from external contamination (microbes and dust) and physical damage, to prolong the shelf life of food products, and to maintain the quality and safety during transport, storage, and distribution [40]. Traditional food packaging is only meant for protection against physical damage and environmental contamination, however, due to the increased concern over food safety and environmental contamination functional food packaging have received huge attention these days. Functional food packaging involves the incorporation of bioactive agents such as antibacterials and antioxidants to the packaging matrix the extend the shelf life of food products [41]. Smart food packaging is another innovative packaging technology that can sense and detect the present condition of food and communicate efficiently to the consumers. For example, during the spoilage of fresh meat and fish products they release volatile amines which causes an increase in their pH level. By using pH sensitive food packaging films, spoilage can be easily detected through color change, help preventing food wastage and health risks associated with consuming spoiled foods [42].

Electrospinning is considered as a simple and economical technique for the fabrication of smart food packaging films [44]. Biopolymer-based electrospun nanofibers with high surface area-to-volume ratio, high porosity, flexibility, biodegradability, and high encapsulation efficiency can effectively carry various bioactive ingredients for food preservation.

Moreover, electrospun nanofiber membranes have excellent oxygen and water vapor barrier properties which helps to inhibit the growth of bacteria and other food spoilage microorganisms. Essential oils are plant-derived compounds with good antioxidant and antimicrobial properties, widely used in food packaging films to extend the shelf-life of perishable food products. Shi et al. [43] fabricated oregano essential oil loaded β-cyclodextrin incorporated PLA/PCL (OEO@βCD/PLA/PCL) nanofiber membranes for blackberry preservation Fig. 3.9. The membrane exhibited good water vapor barrier properties, reducing moisture escape and weight loss during storage. Oregano essential oil was effectively encapsulated within the nanofibers, delaying the fast release of oil and thus making them suitable for long term storage application. Starch-based nanofiber membrane loaded with roselle anthocyanin for the real-time freshness monitoring of shrimp and pork [45], and organic–inorganic hybrid polyvinyl alcohol (PVA) nanofibers with green silver nanoparticles for fruit preservation [46] are few examples of electrospun nanofiber membrane-based smart food packaging systems.

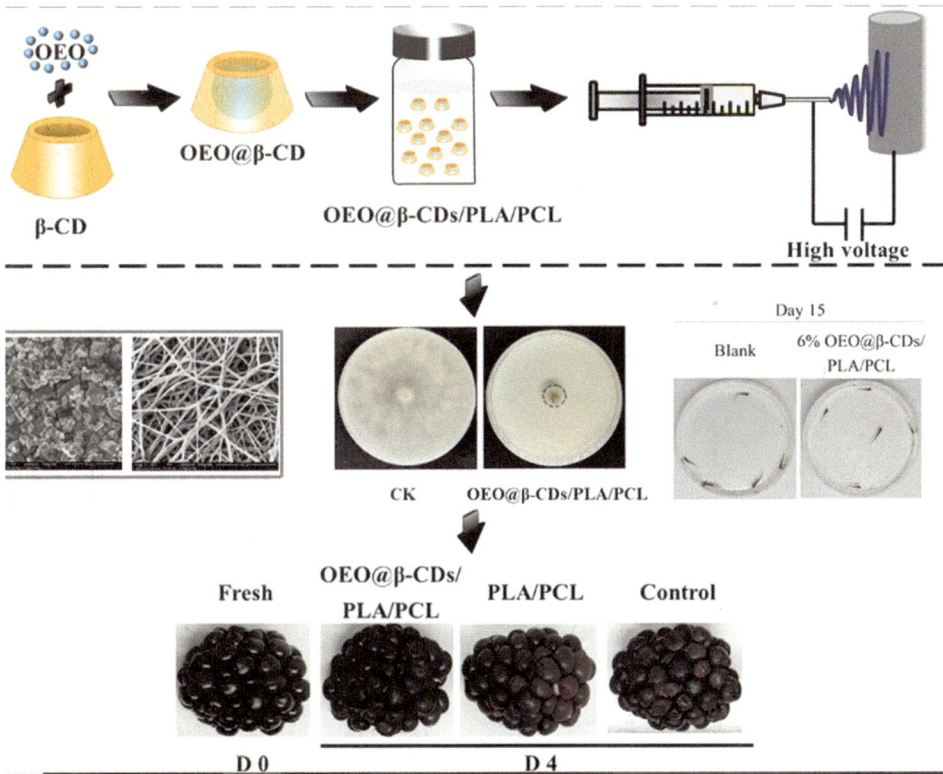

Fig. 3.9 Fabrication and performance of OEO@βCD/PLA/PCL nanofiber membranes for blueberry preservation. Reprinted with permission from Ref. [43]. Copyright Elsevier 2022

Biosensors

Biosensors are analytical devices composed of a biomolecular component capable of sensing analytes (bioreceptors-enzyme, antibody, microorganisms) combined with a physical transducer (optical or electrochemical) capable of sensing various analytes. The bioreceptors attached to the surface of the detection component interact with the target analyte, causing physiochemical changes on its surface, which are subsequently transformed into measurable signals by transducer. Nanotechnology has enabled the design of nanostructures with unique structural features that can significantly improve the performance of biosensing devices. Electrospun nanofibers characterized by a large surface area-to-volume ratio and interconnected porous structures provide numerous active sites which enhances their interactions with target molecules, thereby increasing their sensitivity in detection [47]. Also, the properties of nanofibers are customizable by varying the electrospinning parameters. Their adaptable structure facilitates the integration of other nanomaterials or bioreceptors, further enhancing their performance in fields like medical diagnostics, food analysis, and environmental monitoring. Therefore, electrospun nanofibers provide considerable benefit for biomolecule immobilization, making them well-suited for the development of functional interfaces [48]. Figure 3.10 is a schematic representation of a biosensing device with electrospun nanofibers as the sensing layer.

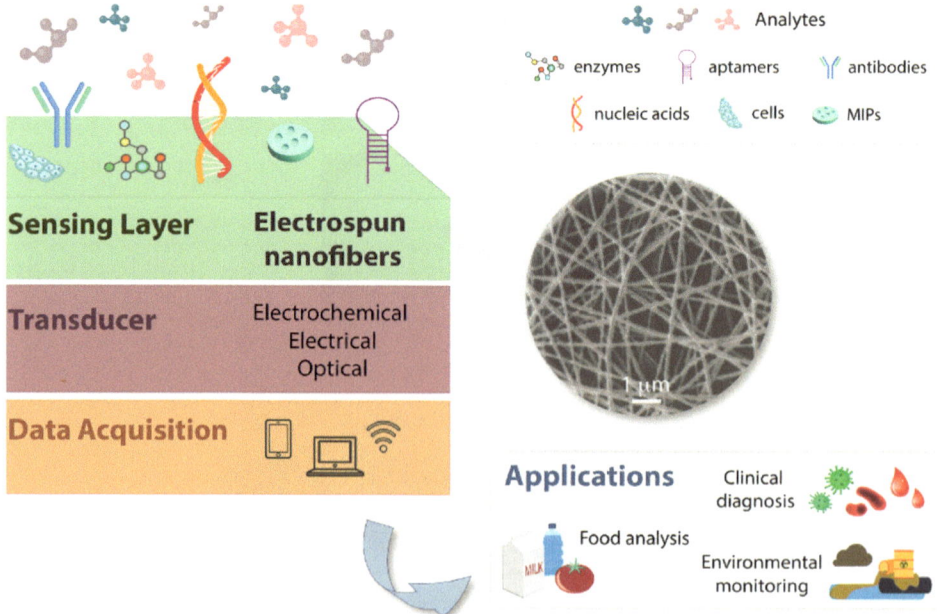

Fig. 3.10 Schematic illustration of biosensing device setup based on electrospun nanofibers. Reprinted with permission from Ref. [48]. Copyright Elsevier 2021

Electropun nanofibers incorporated with metal oxide nanoparticles, carbon particles (carbon nanotubes, graphene), metal organic frame works (MOFs), and conducting polymers are widely used for wearable sensors for health monitoring. Recently, an electrochemical immunosensor wad developed by Kirbay et al. [49] based on electrospun poly-ε-caprolactone (PCL)/poly-$_L$-Lysine (PLL) nanofibers as immobilization matrix for the detection of serum amyloid (SAA). SAA is a protein released from liver in response to acute inflammation and chronic inflammatory conditions. Hence, it serves as a biomarker of inflammation. The immunosensor was fabricated as follows: screen-printed carbon electrode surface was first covered with PCL/PLL nanofibers, followed by the immobilization of serum amyloid A antibody (Anti-SAA). The immunosensor, which integrates the benefits of PCL/PLL nanofibers demonstrated excellent electrochemical performance, high sensitivity, and consistent reproducibility.

References

1. Yadav, T. *et al.* Electrospinning: An Efficient Biopolymer-Based Micro- and Nanofibers Fabrication Technique. in 209–241 (2019). https://doi.org/10.1021/bk-2019-1329.ch010.
2. Al-Abduljabbar, A. & Farooq, I. Electrospun Polymer Nanofibers: Processing, Properties, and Applications. *Polymers (Basel).* **15**, (2023).
3. Haider, A., Haider, S. & Kang, I.-K. A comprehensive review summarizing the effect of electrospinning parameters and potential applications of nanofibers in biomedical and biotechnology. *Arab. J. Chem.* **11**, 1165–1188 (2018).
4. Abdulhussain, R., Adebisi, A., Conway, B. R. & Asare-Addo, K. Electrospun nanofibers: Exploring process parameters, polymer selection, and recent applications in pharmaceuticals and drug delivery. *J. Drug Deliv. Sci. Technol.* **90**, 105156 (2023).
5. Krifa, M. & Yuan, W. Morphology and pore size distribution of electrospun and centrifugal forcespun nylon 6 nanofiber membranes. *Text. Res. J.* **86**, 1294–1306 (2016).
6. Langwald, S. V., Ehrmann, A. & Sabantina, L. Measuring Physical Properties of Electrospun Nanofiber Mats for Different Biomedical Applications. *Membranes (Basel).* **13**, 20–24 (2023).
7. Wang, T. *et al.* Fractal Characteristics of Porosity of Electrospun Nanofiber Membranes. *Math. Probl. Eng.* **2020**, 2503154 (2020).
8. Aghayari, S. The Porosity of Nanofiber Layers. in *Biocomposites* (eds. Elnashar, M. M. M. & Karakuş, S.) (IntechOpen, Rijeka, 2022). https://doi.org/10.5772/intechopen.109104.
9. Juraij, K. *et al.* Polyurethane/multi-walled carbon nanotube electrospun composite membrane for oil/water separation. *J. Appl. Polym. Sci.* **139**, 52117 (2022).
10. Ziel, R., Haus, A. & Tulke, A. Quantification of the pore size distribution (porosity profiles) in microfiltration membranes by SEM, TEM and computer image analysis. *J. Memb. Sci.* **323**, 241–246 (2008).
11. Cheng, T., Li, S., Xu, L. & Ahmed, A. Controllable preparation and formation mechanism of nanofiber membranes with large pore sizes using a modified electrospinning. *Mater. Des.* **178**, 107867 (2019).
12. Ozden, A. *et al.* Chapter 2.28 - Gas Diffusion Layers for PEM Fuel Cells: Ex- and In-Situ Characterization. in *Exergetic, Energetic and Environmental Dimensions* (eds. Dincer, I., Colpan, C. O. & Kizilkan, O.) 695–727 (Academic Press, 2018). https://doi.org/10.1016/B978-0-12-813734-5.00040-8.

13. Rashid, T. U., Gorga, R. E. & Krause, W. E. Mechanical Properties of Electrospun Fibers—A Critical Review. *Adv. Eng. Mater.* **23**, (2021).

14. Han, W.-H. *et al.* Electrospun aligned nanofibers: A review. *Arab. J. Chem.* **15**, 104193 (2022).

15. Huang, Z.-M., Zhang, Y.-Z., Kotaki, M. & Ramakrishna, S. A review on polymer nanofibers by electrospinning and their applications in nanocomposites. *Compos. Sci. Technol.* **63**, 2223–2253 (2003).

16. Matthews, J. A., Wnek, G. E., Simpson, D. G. & Bowlin, G. L. Electrospinning of collagen nanofibers. *Biomacromolecules* **3**, 232–238 (2002).

17. Khajavi, R. & Abbasipour, M. Electrospinning as a versatile method for fabricating coreshell, hollow and porous nanofibers. *Sci. Iran.* **19**, 2029–2034 (2012).

18. Badmus, M., Liu, J., Wang, N., Radacsi, N. & Zhao, Y. Hierarchically electrospun nanofibers and their applications: A review. *Nano Mater. Sci.* **3**, 213–232 (2021).

19. Yoon, J., Yang, H.-S., Lee, B.-S. & Yu, W.-R. Recent Progress in Coaxial Electrospinning: New Parameters, Various Structures, and Wide Applications. *Adv. Mater.* **30**, 1704765 (2018).

20. Han, D. & Steckl, A. J. Coaxial Electrospinning Formation of Complex Polymer Fibers and their Applications. *Chempluschem* **84**, 1453–1497 (2019).

21. Zhang, C., Feng, F. & Zhang, H. Emulsion electrospinning: Fundamentals, food applications and prospects. *Trends Food Sci. Technol.* **80**, 175–186 (2018).

22. Farokhi, M., Mottaghitalab, F., Reis, R. L., Ramakrishna, S. & Kundu, S. C. Functionalized silk fibroin nanofibers as drug carriers: Advantages and challenges. *J. Control. Release* **321**, 324–347 (2020).

23. Zhang, C. *et al.* Core-shell nanofibers electrospun from O/W emulsions stabilized by the mixed monolayer of gelatin-gum Arabic complexes. *Food Hydrocoll.* **107**, 105980 (2020).

24. Qu, H., Wei, S. & Guo, Z. Coaxial electrospun nanostructures and their applications. *J. Mater. Chem. A* **1**, 11513–11528 (2013).

25. Qin, X. 3 - Coaxial electrospinning of nanofibers. in *Electrospun Nanofibers* (ed. Afshari, M.) 41–71 (Woodhead Publishing, 2017). https://doi.org/10.1016/B978-0-08-100907-9.00003-9.

26. Han, D. & Steckl, A. J. Triaxial electrospun nanofiber membranes for controlled dual release of functional molecules. *ACS Appl. Mater. Interfaces* **5**, 8241–8245 (2013).

27. Wang, P., Lv, H., Cao, X., Liu, Y. & Yu, D. G. Recent Progress of the Preparation and Application of Electrospun Porous Nanofibers. *Polymers (Basel).* **15**, (2023).

28. Min, T. *et al.* Novel antimicrobial packaging film based on porous poly(lactic acid) nanofiber and polymeric coating for humidity-controlled release of thyme essential oil. *LWT* **135**, 110034 (2021).

29. Liu, R. *et al.* Progress of Fabrication and Applications of Electrospun Hierarchically Porous Nanofibers. *Adv. Fiber Mater.* **4**, 604–630 (2022).

30. Katsogiannis, K. A. G., Vladisavljević, G. T. & Georgiadou, S. Porous electrospun polycaprolactone (PCL) fibres by phase separation. *Eur. Polym. J.* **69**, 284–295 (2015).

31. Huang, C. & Thomas, N. L. Fabricating porous poly (lactic acid) fi bres via electrospinning. *Eur. Polym. J.* **99**, 464–476 (2018).

32. Nagamine, S., Matsumoto, T., Hikima, Y. & Ohshima, M. Fabrication of porous carbon nanofibers by phosphate-assisted carbonization of electrospun poly(vinyl alcohol) nanofibers. *Mater. Res. Bull.* **79**, 8–13 (2016).

33. Rahmathullah, A. M., Jason Robinette, E., Chen, H., Elabd, Y. A. & Palmese, G. R. Plasma assisted synthesis of hollow nanofibers using electrospun sacrificial templates. *Nucl. Instruments Methods Phys. Res. Sect. B Beam Interact. with Mater. Atoms* **265**, 23–30 (2007).

34. Dalton, P. D., Klee, D. & Möller, M. Electrospinning with dual collection rings. *Polymer (Guildf).* **46**, 611–614 (2005).

35. Kumar, L. *et al.* Hollow Au@TiO2 porous electrospun nanofibers for catalytic applications. *RSC Adv.* **10**, 6592–6602 (2020).

36. Borah, A. R., Hazarika, P., Duarah, R., Goswami, R. & Hazarika, S. Biodegradable Electrospun Membranes for Sustainable Industrial Applications. *ACS Omega* **9**, 11129–11147 (2024).

37. Yan, B., Zhang, Y., Li, Z., Zhou, P. & Mao, Y. Electrospun nanofibrous membrane for biomedical application. *SN Appl. Sci.* (2022) https://doi.org/10.1007/s42452-022-05056-2.

38. Bhardwaj, N. & Kundu, S. C. Electrospinning: A fascinating fiber fabrication technique. *Biotechnol. Adv.* **28**, 325–347 (2010).

39. Wu, X. *et al.* Biomineralized tetramethylpyrazine-loaded PCL/gelatin nanofibrous membrane promotes vascularization and bone regeneration of rat cranium defects. *J. Nanobiotechnology* **21**, 1–21 (2023).

40. Marsh, K. & Bugusu, B. Food Packaging—Roles, Materials, and Environmental Issues. *J. Food Sci.* **72**, R39–R55 (2007).

41. Pascall, M. A., DeAngelo, K., Richards, J. & Arensberg, M. B. Role and Importance of Functional Food Packaging in Specialized Products for Vulnerable Populations: Implications for Innovation and Policy Development for Sustainability. *Foods* **11**, (2022).

42. Wu, X. *et al.* Preparation and characterization of pH-sensitive intelligent packaging films based on cassava starch/polyvinyl alcohol matrices containing Aronia melanocarpa anthocyanins. *LWT* **194**, 115818 (2024).

43. Shi, C. *et al.* Oregano essential oil/β-cyclodextrin inclusion compound polylactic acid/polycaprolactone electrospun nanofibers for active food packaging. *Chem. Eng. J.* **445**, 136746 (2022).

44. Zhang, M., Ahmed, A. & Xu, L. Electrospun Nanofibers for Functional Food Packaging Application. *Materials (Basel).* **16**, (2023).

45. Lv, H. *et al.* Intelligent food tag: A starch-anthocyanin-based pH-sensitive electrospun nanofiber mat for real-time food freshness monitoring. *Int. J. Biol. Macromol.* **256**, 128384 (2024).

46. E, K., K, M., P, B., A, T. S. & I, J. C. R. Biocompatible silver nanoparticles/poly(vinyl alcohol) electrospun nanofibers for potential antimicrobial food packaging applications. *Food Packag. Shelf Life* **21**, 100379 (2019).

47. Zhou, J. *et al.* Electrospun biosensors for biomarker detection. *Colloid Interface Sci. Commun.* **59**, 100767 (2024).

48. Mercante, L. A. *et al.* Nanofibers interfaces for biosensing: Design and applications. *Sensors and Actuators Reports* **3**, 100048 (2021).

49. Ozturk Kirbay, F. & Odaci, D. Electrospun Poly-ε-caprolactone/Poly-l-lysine (PCL/PLL) Nanofibers as an Emergent Material for the Preparation of Electrochemical Immunosensor to Detect Serum Amyloid A. *ACS Appl. Polym. Mater.* **6**, 3778–3786 (2024).

Electrospun Nanofiber Composites

4

Introduction

Electrospun nanofiber-based composites are new category of materials with diverse properties and functionalities. Electrospun nanofibers have exceptional structural features high specific surface area, high porosity, easy processability, light weight, and tunable surface properties. These features makes them useful for diverse applications such as biomedical implants, drug delivery, energy storage, filtration membranes, and sensors [1]. Composites are made of two or more distinct components with varying physical, chemical, and mechanical properties [2]. Nanofiber composites consist of two phases; a matrix phase and a reinforcing phase. Matrix phase also known as continuous phase or primary phase, secures the reinforcement to preserve the intended shape, while reinforcing phase (discontinuous phase) improves the overall characteristics of the composite. Composites generally possess better properties than their individual components, for example high specific strength, high specific modulus, improved resistance to fatigue and corrosion, and cost effectiveness. Despite the advantages of single-phase nanofibers, they have certain drawbacks that limits their application potential; low strength and stiffness, poor thermal stability, poor conductivity and magnetic properties. In electrospun nanofiber composites, the nanofiber can either act as the matrix phase or as the reinforcing phase. Metal nanoparticles, ceramic particles, and nanotubes are the commonly employed reinforcing phases. Depending upon the matrix phase, electrospun nanofiber composites can be classified as polymer matrix composites, ceramic matrix composites, and metal matrix composites [3]. Several studies have revealed the advantages of nanofiber composites over single phase nanofibers, especially for biomedical applications, structural materials, electronics, and optoelectronics field. So, in this chapter we will elaborate on different nanofiber composites, their properties, and key applications.

© The Author(s), under exclusive license to Springer Nature Switzerland AG 2025   51
M. Mathew et al., *Electrospun Porous Nanofibers*, Synthesis Lectures on Emerging
Engineering Technologies, https://doi.org/10.1007/978-3-031-86106-2_4

Polymer Matrix Composites (PMCs)

PMCs are materials composed of reinforcing fillers or particles dispersed in polymer matrix. Among the various polymer composites, fiber reinforced composites are most widely used due to their desired features such as high mechanical strength, light weight, stiffness, and corrosion resistance [4]. Depending upon the reinforcing material used the characteristics of resulting composite can be altered. In PMCs, the polymer matrix transfers stress between the reinforcing elements, providing rigidity and strength, while maintaining lightweight characteristics. With advancements in nanotechnology, research studies are mainly focused on developing nanofillers as reinforcing materials. At the nanoscale, material properties can be significantly enhanced. Electrospun nanofibers, with their nanoscale characteristics can either act as reinforcing filler or as a matrix phase depending on the applications.

Reinforcement of Polymer Matrices

Compared to bulk materials and microfibers, electrospun nanofiber reinforced PMCs have gained significant progress. Their unique size effect, excellent mechanical and thermal properties make them ideal reinforcement for enhancing the performance of polymer matrices. In fiber reinforced composites either continuous (long fibers) or discontinuous (short fiber) fibers can act as reinforcement. If a composite with maximum strength in one direction is required, then long fibers aligned in one direction can be used. Such continuous fiber composites are more stiffer and stronger than discontinuous fiber composites, where the fibers are randomly oriented [5]. Depending on the type of fibers, different fabrication methods such as dip coating, solution casting, in-situ polymerization and electrospinning can be used to fabricate the composite. Due to fiber entanglement issues, fibers cannot be uniformly dispersed in polymer solutions or melt. So dip coating is considered as the best way to disperse electrospun nanofibers in the polymer matrix [6].

Yang et al. [7] reviewed the preparation, properties, and applications of carbon nanofiber (CNF) reinforced composites. CNFs are characterized by high modulus, low density, and excellent conductivity, superior chemical and mechanical stability, hence suitable for wide range of applications like sensors, catalytic systems, sensors, EMI shields, and flexible anodes. Some of the main polymers used as precursors for preparing CNFs are polyacrylonitrile (PAN), polyimide (PI), polyamide (PA), polyvinyl alcohol (PVA), cellulose, and polymethylmethacrylate (PMMA). Pre-oxidation, carbonization, and graphitization are the major steps involved in the preparation of CNFs [8]. CNFs obtained after these steps are thermally stable, but for attaining ideal mechanical strength and conductivity modification is required. Some of the important modification strategies involve plasma treatment, UV treatment, doping with other materials (metal oxides), and functionalization with polymers to enhance their performance. These modified CNFs can then be used as reinforcing agents in composites to improve the mechanical strength, durability, and conductivity. In fiber-reinforced composites, fibers serve as the primary load bearing component that provides strength and stiffness to the composite, while, the matrix binds the fibers together, distributing, and transferring the load uniformly. Chen and colleagues [9] fabricated toughening and rapid self-healing carbon/epoxy composites based on electrospinning thermoplastic polyamide (PA) nanofiber and introduced them into fiber reinforced polymer (FRP) interlamellar region. The laminate consisted of 8 layers of carbon fiber fabric and 7 layers of PA nanofiber as shown in Fig. 4.1a. The entangled nanofiber structure with large surface area enhanced the surface contact with epoxy matrix, providing better interlaminar adhesion. This enhanced interfacial adhesion helps in redirecting or hindering the propagation of any microcracks developed and prevents delamination. The composite attained good toughening, mechanical strength, and tensile modulus compared to the control material without PA fibers.

Enhancement of Mechanical Properties

Electrospun nanofibers can improve the mechanical properties of composites due to their unique structural features. Their high surface area-to-volume ratio can enhance the adhesion between nanofibers and the matrix, thus facilitating more effective stress transfer. By aligning the nanofibers in specified direction, the tensile strength and stiffness of the composite can be improved. Another important feature of electrospun nanofibers is their high aspect ratio (length to diameter) which is hundred times more than microfibers. High aspect ratio prevents matrix deformation. By absorbing the energy during cracks, nanofibers help to improve the composite toughness by acting as a barrier to crack propagation. Additionally, the porous structure of nanofibers facilitates improved matrix impregnation, thus enhancing fiber-matrix integration [10].

Lasenko et al. [11] studied the mechanical properties of nanocomposites reinforced with polyamide (PA6) nanofibers by introducing PA6-oriented nanofibers into epoxy matrix. Epoxy resin is a common matrix material used in composite fabrication due to their adhesiveness, chemical resistance, electrical insulation, and light weight. However, their poor mechanical and thermal properties limit their applications. This can be improved by adding various nanofillers and additives. So here a nanofiber multilayered laminate nanocomposite was fabricated to study the effect of nanofibers on the tensile and thermal characteristics of epoxy matrix. Hand layup method was used to fabricate the composite. Due to the good interface adhesion between epoxy and PA6 nanofibers, the composite exhibited effective stress transmission. Table 4.1 shows the results for tensile test. The results showed an exceptional reinforcing effect with an increase in tensile moduli by 10.58%. The orientation and structure of nanofibers are the main factors that

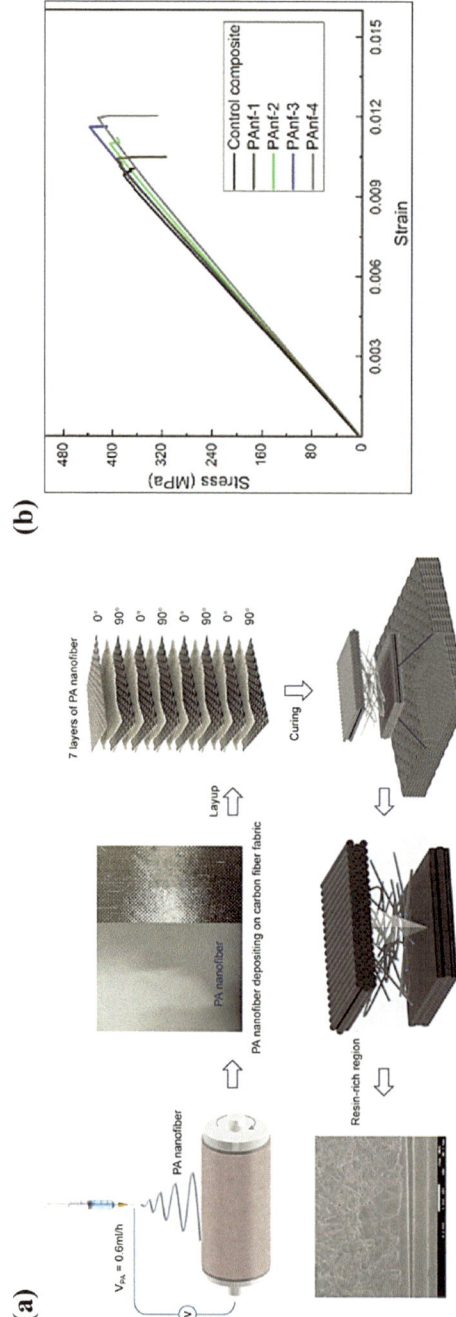

Fig. 4.1 **a** Fabrication of carbon fiber/epoxy composite based on electrospun PA nanofiber, **b** stress–strain curve for composites. Reprinted with permission from Ref. [9]. Copyright John Wiley and Sons 2022

Table 4.1 Tensile test results

Materials	Tensile strength at break, σ (MPa)	Young's modulus, E (MPa)	Elongation at break, ε (%)
PA6 nanofiber	13.18 ± 1.54	3200 ± 15	11.24 ± 1.01
Epoxy	74.45 ± 3.5	2070.4 ± 10	2.32 ± 0.24
Nanocomposite	76.84 ± 4.74	2315.5 ± 19	1.628 ± 0.3

influenced the tensile strength. Strength of composites reinforced with randomly oriented fibers are comparatively lower than fibers oriented in single direction. The rise in elastic modulus can also be related to the fiber orientation, reduced fiber diameter, and the orientation of polymer molecules.

Functionalization for Specific Applications

Electrospun nanofibers can be easily functionalized with variety of natural and synthetic materials like metals, carbon nanoparticles, functional groups, proteins, and peptides for wide range of applications. Surface functionalization can change the physicochemical (wettability, thermal stability, mechanical strength, electrical properties) and biological (cell adhesion, biocompatibility) properties of nanofibers. Surface modification can be achieved using some common methods like plasma treatment, chemical grafting, physical adsorption, layer-by-layer assembly, co-electrospinning of surface-active agents and polymers. Functionalization of polymer nanofibers with various nanocarbons was reviewed by Lee et al. [12]. Nanocarbon functionalized elecrospun composites are receiving enormous attention as next generation functional materials for energy harvesting and storage, ultra-light and ultra-strong composites and biomedical implants. Nanocarbon materials are known for its excellent chemical and thermal properties. Carbon nanotubes, graphene, nanodiamond and carbon quantum dots are some of the common carbon nanomaterials used for reinforcement. Considering the application, polymer-carbon electrospun composites would be a promising scaffold for tissue engineering. Carbon materials can be incorporated into nanofibers either via direct electrospinning or post-electrospinning treatment like dip coating. Several research studies with polymer-nanocarbon composite scaffolds revealed that, in addition to the improvement in mechanical strength an enhanced electrical conductivity and accelerated degradation were also observed.

In a study conducted by Neisiany et al. [13] the mechanical properties of carbon/epoxy composites were improved using functionalized electrospun PAN nanofibers. Initially, glycidyl group from glycidyl methacrylate (GMA) was grafted onto PAN polymer through a free radical reaction. This is followed by the attachment of functionally active epoxy groups, which provide active sited for crosslinking with amines. As shown in Fig. 4.2 the chemical interaction between epoxy groups of PAN-g-GMA and the amine groups in

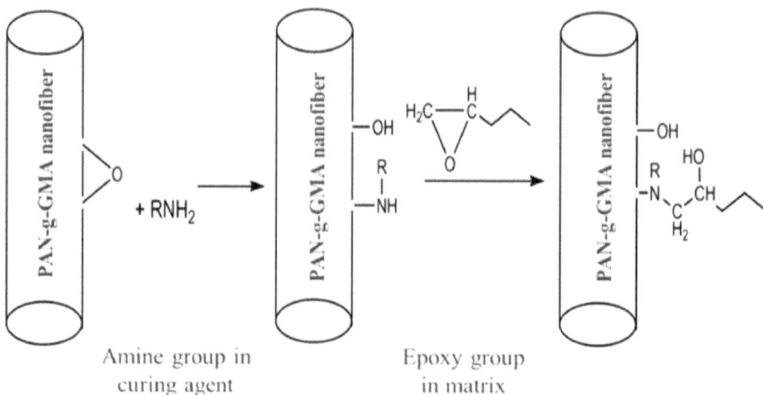

Fig. 4.2 Chemical interaction between epoxy groups of PAN-g-GMA and the amine groups in curing agent. Reprinted with permission from Ref. [13]. Copyright John Wiley and Sons 2017

curing agent strengthen the overall epoxy resin matrix. PAN and PAN-g-GMA nanofibers were electrospun onto unidirectional carbon fabrics, which were then used to fabricate epoxy composites via a wet layup process. Mechanical studies showed that nanofiber functionalization with functional group significantly enhanced the composite's mechanical properties, that is, both in-plane, out-of-plane and impact properties of the composite was benefited from this.

Inorganic Nanofiller Reinforced Composites

Inorganic fillers are inorganic particulate matters having dimensions in the range of nanometers which provide several distinct advantages over conventional fillers [14]. Due to their nano size, they have large surface area to volume ratio allowing better interaction with the polymer matrix, which improves the mechanical and thermal properties of the composite. For optical applications, their reduced light scattering efficiency at optical wavelengths improve their transparency for optical applications and also their physical properties are size dependent. Some of the widely used inorganic fillers for fiber reinforcement include nanoclay, carbon nanotubes, graphene, graphene oxide, layered double hydroxides, and many more [15]. Based on the geometrical structure nanofillers can be classified into three; (1) nanoparticle, (2) nanorod, and (3) nanoplatelet. Since these particles may aggregate into clusters, they are generally exfoliated to obtain a homogeneous dispersion in a nanocomposite. Incorporation of inorganic nanofillers to electrospun fibers improves the mechanical, thermal, and functional properties of nanofibers, making them suitable for high-performance applications. Nanofillers like carbon nanotubes, and graphene can introduce conductivity to electrospun fibers, hence such composites can be used for sensors, electronic, and energy storage devices [16]. Metal oxide nanoparticles (ZnO, TiO_2) are known for its antibacterial and photocatalytic properties, beneficial for applications such as biomedical, filtration, and food packaging [17].

Carbon Nanotubes (CNTs)

Carbon nanotubes are excellent reinforcing material widely used for reinforcing metallic, ceramic, and polymer matrices. CNTs offer excellent mechanical, thermal, and electrical properties to the composite [18]. Also, their nano dimensions and size helps for easy functionalization and tunability. Carbon nanotubes are hollow cylindrical structures made up of hexagonally arranged sp^2 hybridized carbon atoms. Simply it is made by rolling up graphene sheets [19]. Depending upon the number of overlapping layers, CNTs can be classified into: (1) single-walled CNT (SWCNT), (2) double-walled CNT (DWCNT), and (3) multi-walled CNT. The presence of delocalized electrons is responsible for the exceptional conductivity of CNTs [20]. Their high mechanical strength can be attributed to their unique C–C bonding and cylindrical structure. Another important characteristic that influences the optical and electrical properties is the chirality. It has been observed that introduction of CNTs to certain polymer matrices caused the crystallization of the host polymer, enhancing the interaction between CNT and polymer matrix, thus leading to improved mechanical properties [21]. There are various methods available for the synthesis of carbon nanotubes such as arc discharge, laser ablation, chemical vapor deposition and sol gel method. CNTs with various surface morphologies (waved, straight, bent, coiled, beaded) can be fabricated by controlling the synthesis route, catalyst, carbon source, and temperature. Due to these features, CNTs are regarded as outstanding nano-reinforcing materials for electrospun nanofibers. Key factors influencing the performance of CNT-based electrospun composites include the type of CNT, the interfacial interactions with the nanofiber matrix, and the dispersion, distribution, and orientation Of CNTs. CNT-based composites have been used in biomedical fields, nanofiltration membranes, nanosensors, drug delivery, and EMI shielding [22].

Juraij et al. [23] fabricated high strength polyurethane (PU)/multi-walled CNT (MWCNT) electrospun composite membrane for oil/water separation application. There are several studies on the fabrication of membranes with significant strength using CNTs are reinforcing filler. In this study composite membrane with different MWCNT loading were developed. The possible π-π interaction between MWCNT and PU is shown in Fig. 4.3. In comparison to pristine PU membrane, the composite membrane demonstrated exceptional elasticity ($502 \pm 17\%$ elongation) and tensile strength (6.7 MPa). A significant enhancement in thermal stability was also noted with the addition of MWCNT. The optimized composite membrane with 0.2% (w/w) MWCNT showed highest oil sorption capacity and oil flux indicating the potential of mechanically robust electrsopun PU/MWCNT composite membrane for oil/water separation.

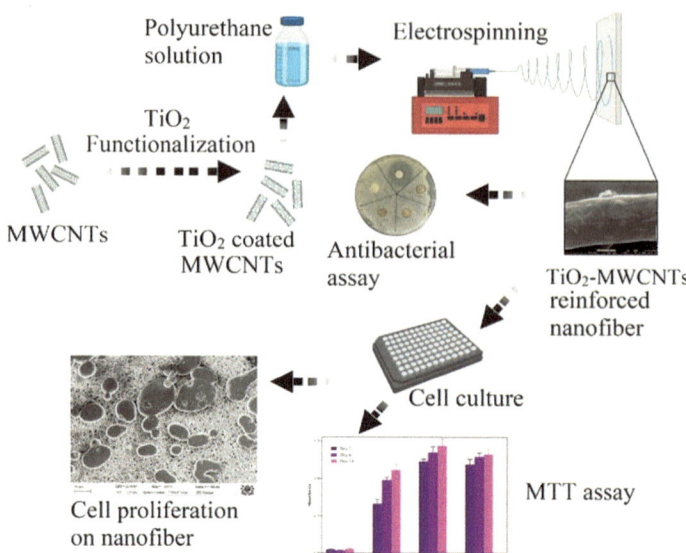

Fig. 4.3 Fabrication and application of TiO_2-MWCNT/Ag nanoparticle reinforced PU nanofiber as tissue engineering scaffold. Reprinted with permission from Ref. [44]. A Copyright Elsevier 2023

Metal-Oxide Nanoparticles

Metal-oxide nanoparticles as reinforcing materials for polymer matrix have received huge attention among many research groups due to their unique structural, electrical, thermal, and mechanical properties [24]. Some of the most common metal-oxide nanoparticles include titanium dioxide (TiO_2), zinc oxide (ZnO), magnesium oxide (MgO), and aluminum oxide (Al_2O_3) [25]. These nanoparticles have high specific surface area and strong interfacial bonding interactions, which significantly enhances the properties of the host polymer matrix. These nanocomposites are widely explored in the areas of photocatalysis, sensors, fuel cells, drug delivery, flame retardants, photodetectors, batteries, and supercapacitors. In nanoparticle reinforced systems, the matrix phase attaches to the nanoparticles via covalent or coordination bonding [26]. The bulk crystal structure of metal-oxides imparts structural rigidity to the nanocomposite. The dispersion of metal-oxide nanoparticles within the polymer matrix is a great challenge due to its aggregation. Hence proper surface modification of the nanoparticles is essential to achieve uniform dispersion within the matrix [27]. Surface modification can be carried out by physical and chemical methods. Physical modification can be achieved through the use of surfactants or macromolecules that adsorb on to the surface of nanoparticles via electrostatic interaction between the polar functional groups of surfactants and the nanoparticle surface. In contrast, chemical modification involves the formation of covalent bonds between nanoparticle surface and a modifying agent [28].

Metal-oxide nanoparticles are commonly used reinforcing materials for enhancing the properties of electrospun nanofibers. Incorporation of these particles enhances the overall functionality of nanofibers making them suitable for multiple applications. Nowadays, biopolymer based electrospun scaffolds are widely utilized for tissue engineering applications, due to their similarity with extracellular matrix, excellent biocompatibility, biodegradability, and cost-effectiveness. However, these scaffolds suffer from poor mechanical strength. One way to overcome this limitation is the incorporation of metaloxide nanoparticles. These nanofillers enable the formation of versatile scaffolds with improved mechanical properties, as well as other features like antimicrobial and anti-inflammatory properties. Alginate is a biopolymer extensively applied in many biomedical fileds like drug delivery, tissue engineering, wound healing, biosensors, and cell encapsulation [29]. In order to enhance the mechanical strength of alginate scaffold Selvi et al. [30] incorporated magnesium oxide (MgO) nanoparticles to reinforce electrospun alginate scaffolds. In this study, spherical MgO nanoparticles were synthesized using a polymer template- assisted ex-situ method and their impact on mechanical, chemical, thermal, and morphological properties of the alginate scaffold was studied. Uniform and bead-free fibers with an average fiber diameter 80–230 nm was obtained. The alginate/MgO nanocomposite scaffold exhibited three times greater tensile strength than alginate-based scaffold. The increased tensile strength is attributed to the effective stress transfer from the matrix phase to the reinforcing fillers, enabled by effective interactions at the filler matrix interphase. The enhanced interfacial interactions restrict the mobility of polymer chains providing structural rigidity to the scaffold. Also, the thermal stability of alginate/ MgO scaffold was higher than that of alginate-based scaffold, which is due to the high thermal stability of MgO nanoparticles (decomposition temperature = 2800 °C). Thus, these findings suggest the suitability of MgO nanoparticle reinforced alginate scaffold for tissue engineering applications.

Hybrid Nanocomposites

Hybrid nanocomposites are materials synthesized by dispersing inorganic nanoparticles into a macroscopic organic matrix [31]. The combination of organic and inorganic materials offers several advantages by utilizing the unique properties of both materials. Generally, organic matrices like polymers provides light weight, low production cost, and large surface area, while enhancing the optical and electrical properties of composites [32]. On the other hand, inorganic particles enhance the mechanical, thermal, and chemical properties making these composites ideal for energy and environmental applications. Mainly there are two types of hybrid composites: polymer-based hybrid composite and hybrid nanoparticles. As already discussed in Sect. 4.1 polymer composites are made by incorporating reinforcing materials into polymer matrix. Hybrid nanoparticles are made by a synergistic combination of two or more components at the nanoscale. Core–shell structures are examples for such kind of composites [33]. In hybrid nanoparticles the inorganic core is covered with organic shell through chemical interactions. The core material provides its unique properties like high conductivity, magnetic, and catalytic properties, while the shell material offers durability, specificity, and improved surface properties [34].

Electrospun nanofiber-based hybrid composites can be fabricated by the incorporation of two or more nanofillers to the polymer solution. An antibacterial hybrid polyacrylonitrile (PAN) nanofiber composites were fabricated by adding Ag, CuO, and ZnO nanoparticles as bactericides by a group of researcher's [35]. PAN is a synthetic polymer widely used for air filtration applications due to its high mechanical strength. However, PAN membranes are susceptible to rapid fouling by bacteria and proteins during filtration due to their hydrophobicity. This might cause severe issues like secondar pollution, lower productivity, and shorter life span. To overcome this, the membrane was functionalized with inorganic metal and metal oxides with excellent antibacterial properties. Owing to the combined effect, the PAN nanofiber membrane integrated with metal oxides or metal ions demonstrated significantly improved antibacterial performance in the hybrid composite compared to the pure PAN nanofiber membrane.

Applications of Nanofiber Composites

Electrospun nanofibers have attracted considerable attention for various biomedical applications, as structural materials, filtration membranes, and sensors owing to their distinctive structural and functional characteristics. Despite their numerous advantages like high surface area-to-volume ratio, high porosity, and high aspect ratio, certain applications demand the use of nanofiber composites. Nanofiber composites possess excellent structural and tunable functional properties compared to single phase nanofibers [36, 37]. This section provides a discussion of some key applications of nanofiber composites.

Structural Materials

Nanofiber composites with improved properties such as excellent mechanical strength, porous architecture, tailorable porosity, and tunable functional properties can serve as high potential structural materials. The selection of a particular material as a structural material is based on its various properties. However, stiffness and strength are the two primary characteristics that determine their ability to perform under mechanical loads and maintain their shape and functionality [38]. A material's strength can be defined as how much stress it can withstand before it deforms permanently and stiffness is how much the material deforms under a given load. Due to their exceptional properties, structural materials are widely used in transportation, aerospace, biomedical implants, civil infrastructure, and sporting goods [39].

In a study by Lee et al. [40] electrospun carbon nanofibers containing multi-walled carbon nanotubes (CNF-MWCNT) were developed as electrodes for structural supercapacitors. As a structural material, carbon-fiber reinforced polymers (CFRP) are recognized for its light-weight, high specific strength, stiffness, and corrosion resistance. These properties make them ideal for use in load bearing structures. In traditional energy storage devices, the energy storage device (like supercapacitors and batteries) and the structural materials are separate. Batteries do not contribute to the structure's strength and increases the overall weight of the system. So, in this study a novel CFRP composite have been developed which serves dual function: bearing mechanical load and storing energy at the same time. In this structural supercapacitor, carbon nanofibers fabricated via electrospinning was used as the electrode material due to its high specific surface area. To enhance the conductivity, nanofibers were modified with MWCNT. According to the results, the structural supercapacitor demonstrated highest level of multifunctionality.

Tissue Engineering Scaffolds

Electrospun nanofiber composites have outstanding properties suitable for tissue engineering scaffolds such as high specific surface area, high porosity, mimicking extracellular matrix (ECM), and the ability to tune their surface functionality. Recent advances in electrospinning allows for the precise modification of nano surfaces in tissue engineering scaffolds, allowing the controlled adsorption of proteins and formation of protein layers that resemble the ECM. Thus, these modified tissue engineering scaffolds play a pivotal role in cell adhesion, proliferation, nutrient transport, and signal transduction making electrospun fibers valuable in tissue engineering applications [41, 42].

In a recent literature review by Zhang et al. [43] they discussed the fabrication and application of nanofiber/hydrogel composite scaffolds for skin, blood vessel, nerve and other tissue engineering applications. Hydrogels are one of the most commonly used tissue engineering scaffolds because of their high-water holding capacity which is similar to those of soft tissues. However, they cannot mimic the complex three-dimensional fibrous structure of ECM. So inorder to overcome this issue, hydrogels are combined with nanofibers that can mimic the ECM. Thus, integrating nanofibers with hydrogels can produce a new composite material that combines the functional properties of hydrogel like good biocompatibility, and high-water content, with structural benefits of nanofiber. There are several strategies for fabricating nanofiber/hydrogel composite. A nanofiber/hydrogel composite can be created by coating a nanofibrous membrane with a hydrogel precursor or embedding electrospun fibers within the hydrogel matrix. 3D structures that can mimic natural tissues can be obtained by layer-by-layer assembly which has received huge attention among the researchers.

In another study, polyurethane-based (PU) bone regenerative nanofibers reinforced with titanium dioxide (TiO_2)-MWCNT and silver (Ag) nanoparticles were fabricated as a multifunctional scaffold [44]. PU is an excellent material widely used in surgical instruments, biomedical implants, and health care beddings. PU possess several key properties like it is bioactive, hence it can interact with biological tissues to support healing, and particularly used as bone cement. Its osteoinductive properties stimulate bone growth making it a perfect material for bone repair and regeneration. Here, TiO_2 functionalized MWCNT was used to improve the mechanical strength of PU scaffold in addition to their inherent osteogenic property. Pure PU nanofibers showed only minimum tensile strength of 2.1 MPa and the composite membrane exhibited a maximum tensile strength of 7.1 MPa with increase in concentration of TiO_2-MWCNT. The incorporation of Ag nanoparticles improves the hydrophilic character and controls bacterial growth near the implantation site. The overall scheme illustrating the fabrication is given in Fig. 4.3.

Electronics and Optoelectronics

Electrospun nanofibers have attained considerable interest in the field of electronics and optoelectronics owing to their superior properties such as high surface area, tunable porosity, flexibility, conductivity, and the ability to integrate functional materials [45]. These fibers are used in batteries, supercapacitors, diodes, solar cells, sensors, and electrochromic devices. Depending on the device requirements, they can be used as single filaments, as fibrous webs or as aligned fiber arrays [46].

Dye sensitized solar cells (DSSC) are considered as third generation highly efficient solar cells. DSSCs come under thin-film solar cell with lots of advantages like low-cost, non-toxicity, simple preparation and ease of production. These devices generate electricity through illuminated organic dyes in electrochemical cells. The major components of a DSSC are: photoanode, dye sensitizer layer, electrolyte, and counter electrode [47]. A photoanode typically consists of a metal oxide semiconductor layer, such as TiO_2, which is deposited on to a transparent conductive glass substrate. The dye sensitizer layer is covalently attached to the photoanode for absorbing light to generate electric current through photon excitation. The redox electrolyte provides a transport path for electron transfer from the counter electrode to the oxidized dye. That is finally, a counter electrode to collect the electrons from the external circuit and dispersing them in electrolyte. Despite of its huge advantages, its low energy conversion efficiency limits its applications. One approach to overcome this limitation is the fabrication of components using electrospinning [48]. Leakage of liquid electrolytes is a major issue that diminishes the efficiency of most energy storage system. Replacing organic liquid electrolyte with gel polymer electrolyte is better option. Gel polymer electrolytes are formed by embedding a liquid electrolyte within a polymer matrix. In a study by Zhao et al. [49] polyacrylonitrile (PAN)-based electrospun composite fibrous membranes incorporated with ceramic

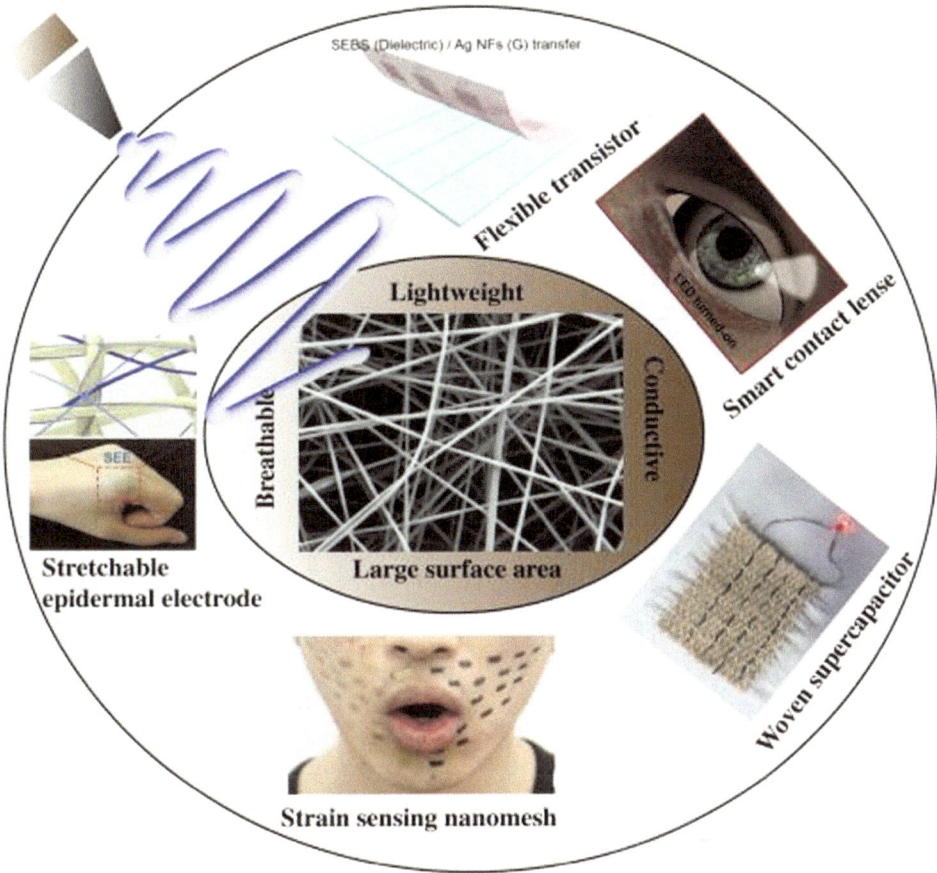

Fig. 4.4 Wearable devices based on electrospun nanofibers [50]

filler SiO_2 as polymer electrolyte for enhanced DSSC efficiency. The composite membrane exhibited a three dimensional, interconnected porous network with homogeneously distributed SiO_2 nanoparticles. The highly porous structure allows for the encapsulation of large amount of electrolyte with high ionic conductivities. The addition of SiO_2 not only lowered the electrolyte resistance, but also improved the charge transfer. As a result, DSSC with fibrous composite membrane containing 10 wt% SiO_2 attained high conversion efficiency of 7.85% at 100 mWcm^{-2} and exhibited long term stability compared to liquid electrolyte-based DSSCs.

Electrospun nanofiber-based wearables is another progressing field [50]. These are miniature electronic devices that can be directly worn on human skin for various sensing applications. Figure 4.4 represents wearable devices made using electrospinning. Electrospun nanofibers are ideal on-skin biosensors suitable for personalized health monitoring owing to their high porosity, breathability, light weight, and high specific surface area. They are highly permeable to gas and water vapor, allowing effective sensing at the skin interface. Functionalizing these fibers enables them to detect various bio-signals with advanced healthcare applications.

Conclusion

Electrospinning is a simple and versatile technique used for the fabrication of ultrathin fibers with unique structural properties like large specific surface area, high porosity, easy processability, and tunable surface properties. Electrospinning allows for well-defined control over fiber morphology and composition, enhancing mechanical, thermal, and barrier properties. Due to these features, electrospun nanofibers have been considered as excellent reinforcements in composites. Several factors influence the property of nanofiber reinforced composites such as fiver diameter, fiber alignment, fiber/matrix interfacial interaction, and so on. Since surface functionalization is possible, modifications with bioactive agents provide targeted functionalities. Thus, nanofiber-based composites are potential candidates capable of multifunctional material solutions in various industrial applications.

References

1. Agarwal, S., Greiner, A. & Wendorff, J. H. Electrospinning of Manmade and Biopolymer Nanofibers—Progress in Techniques, Materials, and Applications. *Adv. Funct. Mater.* **19**, 2863–2879 (2009).
2. Sharma, A. K., Bhandari, R., Aherwar, A. & Rimašauskienė, R. Matrix materials used in composites: A comprehensive study. *Mater. Today Proc.* **21**, 1559–1562 (2020).
3. Ramalingam, M. & Ramakrishna, S. *Introduction to Nanofiber Composites. Nanofiber Composites for Biomedical Applications* (Elsevier Ltd, 2017). https://doi.org/10.1016/B978-0-08-100 173-8.00001-6.
4. Shakil, U. A., Hassan, S. B. A., Yahya, M. Y. & Nauman, S. Mechanical properties of electrospun nanofiber reinforced/interleaved epoxy matrix composites—A review. *Polym. Compos.* **41**, 2288–2315 (2020).
5. Böhlke, T. *et al.* Continuous—Discontinuous Fiber-Reinforced Polymers. in *Continuous-Discontinuous Fiber-Reinforced Polymers* (eds. Böhlke, T. et al.) I–XXVIII (Hanser, 2019). https://doi.org/10.3139/9781569906934.fm.
6. Jiang, S. *et al.* Electrospun nanofiber reinforced composites: A review. *Polym. Chem.* **9**, 2685–2720 (2018).
7. Yang, X., Chen, Y., Zhang, C., Duan, G. & Jiang, S. Electrospun carbon nanofibers and their reinforced composites: Preparation, modification, applications, and perspectives. *Compos. Part B Eng.* **249**, 110386 (2023).
8. Arshad, S. N., Naraghi, M. & Chasiotis, I. Strong carbon nanofibers from electrospun polyacrylonitrile. *Carbon N. Y.* **49**, 1710–1719 (2011).

9. Chen, B., Cai, H., Mao, C., Gan, Y. & Wei, Y. Toughening and rapid self-healing for carbon fiber/epoxy composites based on electrospinning thermoplastic polyamide nanofiber. *Polym. Compos.* **43**, 3124–3135 (2022).

10. Mohammadzadehmoghadam, S., Dong, Y. & Jeffery Davies, I. Recent progress in electrospun nanofibers: Reinforcement effect and mechanical performance. *J. Polym. Sci. Part B Polym. Phys.* **53**, 1171–1212 (2015).

11. Lasenko, I. *et al.* The Mechanical Properties of Nanocomposites Reinforced with PA6 Electrospun Nanofibers. *Polymers (Basel).* **15**, (2023).

12. Lee, J. K. Y. *et al.* Polymer-based composites by electrospinning: Preparation & functionalization with nanocarbons. *Prog. Polym. Sci.* **86**, 40–84 (2018).

13. Neisiany, R. E., Khorasani, S. N., Naeimirad, M., Lee, J. K. Y. & Ramakrishna, S. Improving Mechanical Properties of Carbon/Epoxy Composite by Incorporating Functionalized Electrospun Polyacrylonitrile Nanofibers. *Macromol. Mater. Eng.* **302**, 1600551 (2017).

14. Imai, Y. Inorganic Nano-fillers for Polymers. in *Encyclopedia of Polymeric Nanomaterials* (eds. Kobayashi, S. & Müllen, K.) 1–7 (Springer Berlin Heidelberg, Berlin, Heidelberg, 2021). https://doi.org/10.1007/978-3-642-36199-9_353-1.

15. George, J. & Ishida, H. A review on the very high nanofiller-content nanocomposites: Their preparation methods and properties with high aspect ratio fillers. *Prog. Polym. Sci.* **86**, 1–39 (2018).

16. Ludwig, T. *et al.* Inorganic Nanofibers by Electrospinning Techniques and Their Application in Energy Conversion and Storage Systems. *Semicond. Semimetals* **98**, 1–70 (2018).

17. Deshmukh, K., Sankaran, S., Basheer Ahamed, M. & Khadheer Pasha, S. K. Biomedical Applications of Electrospun Polymer Composite Nanofibres. in *Polymer Nanocomposites in Biomedical Engineering* (eds. Sadasivuni, K. K., Ponnamma, D., Rajan, M., Ahmed, B. & Al-Maadeed, M. A. S. A.) 111–165 (Springer International Publishing, Cham, 2019). https://doi.org/10.1007/978-3-030-04741-2_5.

18. Rani, P., Ahamed, M. B. & Deshmukh, K. Carbon Nanotubes Embedded in Polymer Nanofibers by Electrospinning. in *Handbook of Carbon Nanotubes* (eds. Abraham, J., Thomas, S. & Kalarikkal, N.) 1–35 (Springer International Publishing, Cham, 2020). https://doi.org/10.1007/978-3-319-70614-6_12-1.

19. Kausar, A. 2 - Carbonaceous nanofillers in polymer matrix. in *Polymeric Nanocomposites with Carbonaceous Nanofillers for Aerospace Applications* (ed. Kausar, A.) 23–53 (Woodhead Publishing, 2023). https://doi.org/10.1016/B978-0-323-99657-0.00009-0.

20. Mondal, S. 11 - Carbon nanotube-reinforced polymer nanocomposite for biomedical applications. in *Green Biocomposites for Biomedical Engineering* (eds. Hoque, M. E., Sharif, A. & Jawaid, M.) 265–283 (Woodhead Publishing, 2021). https://doi.org/10.1016/B978-0-12-821553-1.00013-2.

21. Naebe, M., Lin, T. & Wang, X. Carbon Nanotubes Reinforced Electrospun Polymer Nanofibres. in *Nanofibers* (ed. Kumar, A.) (IntechOpen, Rijeka, 2010). https://doi.org/10.5772/8160.

22. Mehdipour-Ataei, S. & Aram, E. 18 - High-end applications of unsaturated polyester composites. in *Applications of Unsaturated Polyester Resins* (eds. Thomas, S. & Chirayil, C. J.) 421–439 (Elsevier, 2023). https://doi.org/10.1016/B978-0-323-99466-8.00009-5.

23. Juraij, K. *et al.* Polyurethane/multi-walled carbon nanotube electrospun composite membrane for oil/water separation. *J. Appl. Polym. Sci.* **139**, 52117 (2022).

24. Wawrzyńczak, A., Chudzińska, J. & Feliczak-Guzik, A. Metal and Metal Oxides Nanoparticles as Nanofillers for Biodegradable Polymers. *ChemPhysChem* **25**, e202300823 (2024).

25. Khalil, M., Jan, B. M., Tong, C. W. & Berawi, M. A. Advanced nanomaterials in oil and gas industry: Design, application and challenges. *Appl. Energy* **191**, 287–310 (2017).

26. Zadehnazari, A. Metal oxide/polymer nanocomposites: A review on recent advances in fabrication and applications. *Polym. Technol. Mater.* **62**, 655–700 (2023).
27. Anupama Hiremath Amar A Murthy, S. T. & N, B. K. Nanoparticles Filled Polymer Nanocomposites: A Technological Review. *Cogent Eng.* **8**, 1991229 (2021).
28. Ahangaran, F. & Navarchian, A. H. Recent advances in chemical surface modification of metal oxide nanoparticles with silane coupling agents: A review. *Adv. Colloid Interface Sci.* **286**, 102298 (2020).
29. Lee, K. Y. & Mooney, D. J. Alginate: Properties and biomedical applications. *Prog. Polym. Sci.* **37**, 106–126 (2012).
30. De Silva, R. T. *et al.* Magnesium Oxide Nanoparticles Reinforced Electrospun Alginate-Based Nanofibrous Scaffolds with Improved Physical Properties. *Int. J. Biomater.* **2017**, 1391298 (2017).
31. Palanisamy, P., Chavali, M., Kumar, E. M. & Etika, K. C. *Hybrid Nanocomposites and Their Potential Applications in the Field of Nanosensors/Gas and Biosensors. Nanofabrication for Smart Nanosensor Applications* (Elsevier Inc., 2020). https://doi.org/10.1016/B978-0-12-820702-4.00011-8.
32. Nguyen, T.-P. & Yang, S.-H. 19 - Hybrid materials based on polymer nanocomposites for environmental applications. in *Polymer-based Nanocomposites for Energy and Environmental Applications* (eds. Jawaid, M. & Khan, M. M.) 507–551 (Woodhead Publishing, 2018). https://doi.org/10.1016/B978-0-08-102262-7.00019-2.
33. Singh, P. *et al.* Hybrid silver nanoparticles: Modes of synthesis and various biomedical applications. *Electron* **2**, e22 (2024).
34. Ghosh Chaudhuri, R. & Paria, S. Core/Shell Nanoparticles: Classes, Properties, Synthesis Mechanisms, Characterization, and Applications. *Chem. Rev.* **112**, 2373–2433 (2012).
35. Shalaby, T., Hamad, H., Ibrahim, E., Mahmoud, O. & Al-Oufy, A. Electrospun nanofibers hybrid composites membranes for highly efficient antibacterial activity. *Ecotoxicol. Environ. Saf.* **162**, 354–364 (2018).
36. Ramalingam, M. & Ramakrishna, S. 1 - Introduction to nanofiber composites. in *Nanofiber Composites for Biomedical Applications* (eds. Ramalingam, M. & Ramakrishna, S.) 3–29 (Woodhead Publishing, 2017). https://doi.org/10.1016/B978-0-08-100173-8.00001-6.
37. Sahay, R. *et al.* Electrospun composite nanofibers and their multifaceted applications. *J. Mater. Chem.* **22**, 12953–12971 (2012).
38. Anwar, N. & Najam, F. A. Chapter One - Structures and Structural Design. in *Structural Cross Sections* (eds. Anwar, N. & Najam, F. A.) 1–37 (Butterworth-Heinemann, 2017). https://doi.org/10.1016/B978-0-12-804443-8.00001-4.
39. Chung, D. D. L. Sensing Materials: Self-Sensing Materials. in *Encyclopedia of Sensors and Biosensors (First Edition)* (ed. Narayan, R.) 196–203 (Elsevier, Oxford, 2023). https://doi.org/10.1016/B978-0-12-822548-6.00004-2.
40. Lee, D. *et al.* Structural energy storage system using electrospun carbon nanofibers with carbon nanotubes. *Polym. Compos.* **45**, 2127–2139 (2024).
41. Elena Ekrami Mahvash Khodabandeh Shahraky, M. M. M. S. M. & Shariati, P. Biomedical applications of electrospun nanofibers in industrial world: a review. *Int. J. Polym. Mater. Polym. Biomater.* **72**, 561–575 (2023).
42. Yan, B., Zhang, Y., Li, Z., Zhou, P. & Mao, Y. Electrospun nanofibrous membrane for biomedical application. *SN Appl. Sci.* (2022) https://doi.org/10.1007/s42452-022-05056-2.
43. Zhang, M. *et al.* Electrospun nanofiber/hydrogel composite materials and their tissue engineering applications. *J. Mater. Sci. Technol.* **162**, 157–178 (2023).

44. Umair Wani, T. *et al.* Titanium dioxide functionalized multi-walled carbon nanotubes, and silver nanoparticles reinforced polyurethane nanofibers as a novel scaffold for tissue engineering applications. *J. Ind. Eng. Chem.* **121**, 200–214 (2023).

45. Wang, Y., Yokota, T. & Someya, T. Electrospun nanofiber-based soft electronics. *NPG Asia Mater.* **13**, (2021).

46. Fang, J., Shao, H., Niu, H. & Lin, T. Applications of Electrospun Nanofibers for Electronic Devices. in *Handbook of Smart Textiles* (ed. Tao, X.) 617–652 (Springer Singapore, Singapore, 2015). https://doi.org/10.1007/978-981-4451-45-1_32.

47. Gong, J., Sumathy, K., Qiao, Q. & Zhou, Z. Review on dye-sensitized solar cells (DSSCs): Advanced techniques and research trends. *Renew. Sustain. Energy Rev.* **68**, 234–246 (2017).

48. López-Covarrubias, J. G., Soto-Muñoz, L., Iglesias, A. L. & Villarreal-Gómez, L. J. Electrospun nanofibers applied to dye solar sensitive cells: A review. *Materials (Basel).* **12**, 1–18 (2019).

49. Zhao, J., Jo, S.-G. & Kim, D.-W. Photovoltaic performance of dye-sensitized solar cells assembled with electrospun polyacrylonitrile/silica-based fibrous composite membranes. *Electrochim. Acta* **142**, 261–267 (2014).

50. Das, R., Zeng, W., Asci, C., Del-Rio-Ruiz, R. & Sonkusale, S. Recent progress in electrospun nanomaterials for wearables. *APL Bioeng.* **6**, 45–49 (2022).